International Conference on Materials Engineering and Characterization 2019

IOP Conference Series: Materials Science and Engineering
Volume 574

Kanchipuram, India
8 – 10 May 2019

ISBN: 978-1-5108-9453-2
ISSN: 1757-8981

Printed from e-media with permission by:

Curran Associates, Inc.
57 Morehouse Lane
Red Hook, NY 12571

Some format issues inherent in the e-media version may also appear in this print version.

Institute of Physics
Dirac House, Temple Back
Bristol BS1 6BE UK

Phone: 44 1 17 929 7481
Fax: 44 1 17 920 0979

techtracking@iop.org

Additional copies of this publication are available from:

Curran Associates, Inc.
57 Morehouse Lane
Red Hook, NY 12571 USA
Phone: 845-758-0400
Fax: 845-758-2633
Email: curran@proceedings.com
Web: www.proceedings.com

Table of contents

Volume 574

International conference on Materials Engineering and characterization

8–10 May 2019, Adhi College of Engineering and Technology, Kanchipuram, Tamil Nadu, India

Accepted papers received: 24 June 2019
Published online: 12 September 2019

Preface

Papers

Material Science

Preface

International conference on Materials Engineering and Characterization was held between May 8-10, 2019 at Adhi College of Engineering and Technology, Sankarapuram, Kancheepuram, Tamil Nadu, India.

The main objective of the conference is to promote the innovative works attempted in the core and diversified fields of materials engineering for different applications. Materials engineering is one of very most important interdisciplinary fields that focuses on studies fundamental physical and chemical basis in order to develop new materials and compounds through manipulating atoms to form desired structures and properties. Any revolutionary changes are always through the breakthrough in new materials.

With this in mind, ICMECh'19 had provided a very important platform to not only materials scientists and engineers but also physicists and chemists to exchange their achievements and to discuss future directions in developing new, resilient and cost-effective materials. It created a so rare opportunity for participants to mingle together for future collaborations in research and development.

Around 100 authors represented from four countries presented their works at the forum of ICMECh'19. The forum comprised 4 plenary and 8 Invited talks, 80 oral presentations and 60 poster presentations covering polymers, ceramics and nanomaterials, nanofluids, natural fiber reinforced composites, hybrid metal matrix composites, synthesis of multifunctional nanomaterials, phase change materials, friction stirrer welding, thin film coatings, tribology, crystals and biomaterials.

We are grateful to the advisory committees for their valuable suggestions and contributions in organizing the mammoth scientific event. Organizing committee sincerely acknowledges the support and continuous encouragement from delegates, experts and participants from various reputed institutions and research centers. Also, it sincerely thanks all the experts who presided during the oral and the poster presentations. A cultural visit to Kancheepuram and Mamallapuram (tourist places) was arranged and with the help of a professional guide, the highlights of the ancient temples and monuments were explained to the participants.

I would like to convey my gratitude to IOP publications for accepting to publish the research findings in IOP Conference Series: Materials Science and Engineering. The first edition of the ICMECH'19 Proceedings includes 23 articles eventually accepted after the peer review process accomplished on them. The organizers sincerely thank to all the reviewers for making their efforts to suggest the authors, so as to improve the quality of the manuscripts before publishing them.

Regards

Prof. I.Saravanan

(Convener-ICMECH)

IOP Conf. Series: Materials Science and Engineering **574** (2019) 011002 doi:10.1088/1757-899X/574/1/011002

Peer review statement

All papers published in this volume of *IOP Conference Series: Materials Science and Engineering* have been peer reviewed through processes administered by the proceedings Editors. Reviews were conducted by expert referees to the professional and scientific standards expected of a proceedings journal published by IOP Publishing.

Experimental Investigation of Aluminum Graphite and MOS2 Composite Materials

Prakash Malaiyappan[a],S.Vignesh[b]
Assistant Professor[a], Saveetha School of Engineering, Chennai, Tamil Nadu.
Student[b], Department of Mechanical Engineering, Saveetha School of Engineering, Chennai, Tamil Nadu, India.
Corresponding author: prakashvictorz@gmail.com

Abstract. Metal Matrix Composites (MMCs) have light weight, high quality, firmness and display more prominent protection from consumption, oxidation, and wear. In this investigation half and half Aluminum network composite was created through blend throwing course. Molybdenum disulfide (MoS2) and graphite (Gr) particles were utilized as fortification stages for the present investigation. Three grinding material examples were created dependent on a trial detailing, and vol. % graphite + vol. % MoS2, individually.

1. INTODUCTION

This content deals with the composition of three materials to change its behavior by increasing wear, stiffness and ability against corrosion. In this aluminum is used as major material because of its low weight, high strength, excellent corrosion resistance, good thermal and electrical conductivity to increase the property of aluminum graphite and molybdenum disulfide as shown Fig. 1 is used.

1.1. Structure

After iron, aluminum is presently the second most generally utilized metal on the planet. The properties of aluminum include: low thickness and along these lines low weight, high quality, unrivaled flexibility, simple machining, superb erosion opposition and great warm and electrical conductivity are among aluminum's most significant properties. Aluminum is additionally extremely simple to reuse [6] Graphite's Non-metallic properties incorporate idleness, high warm opposition and lubricity. A portion of the real end employments of graphite are in high-temperature ointments. Graphite is a layer grid lamella precious stone structure where the bonds between the carbon iotas in the gem structure of the layer are more grounded than the carbon bonds between layers. Graphite is contained carbon and water vapor. Every carbon iota is clung to three other encompassing carbon particles. The level rings of carbon molecules are reinforced into hexagonal structures. The hexagonal carbon structure is alluded to as a benzene ring.

MoS2 has a hexagonal crystalline structure. The natural property of simple shear happens at the interface between the sulfur particles. The connection between layers gives an impact like what an individual may involvement if endeavoring to move over a story totally secured with fresh out of the plastic new playing a game of cards. Each playing card slides effectively concerning every individual layer. The molecule size and film thickness are significant parameters that ought to be coordinated to the surface harshness of the greased up part

1.2. CompositeMaterials:

For composite material selection of Matrix and reinforcement are of prime importance. Aluminum is a synthetic component in the boron bunch with image Al and nuclear number 13. It is a shimmering white, delicate, flexible metal. Aluminum is the third most inexhaustible component and the most plenteous metal in the Earth's outside layer. It makes up about 8% by weight of the Earth's strong surface. Aluminum metal is so artificially responsive that local examples are uncommon and constrained to outrageous lessening conditions. Rather, it is discovered joined in more than 270 distinct minerals. The main mineral of aluminum is bauxite. Aluminum is wonderful for the metal's low thickness and for its capacity to oppose erosion because of the marvel of passivation. Basic segments produced using aluminum and its composites are essential to the aeronautic trade and are significant in different regions of transportation and auxiliary materials. The most helpful mixes of aluminum, at any rate on a weight prize, are the oxides and sulfates. In spite of its pervasiveness in the earth, no known type of life utilizes aluminum salts metabolically. With regards to its inescapability, aluminum is all around endured by plants and creatures. Inferable from their predominance, potential useful (or something else) organic jobs of aluminum mixes are of proceeding with intrigue.

1.2.1. Chemical composition of pure aluminum:

Si	**Fe**	**Ti**	**V**	**Cu**	**Mn**	**Al**	**others**
0.07%	0.34%	0.001%	0.008%	0.001%	0.004%	99.55%	0.002%

Particles of MoS2 and graphite are used as reinforcement.Addition of graphite particle results in low friction of composite as it is good dry lubricant hence reduces wear and abrasion.Molybdenum disulfide is similar to graphite. It is widely used as a solid lubricant because of its low friction properties and robustness.

2. FABRICATION PROCESS

In this project we make the Aluminum Rod in different materials composite. The materials are Aluminum, Graphite and MoS2. In this Rod the major Material was Aluminum and some percentage of Graphite and MoS2. Three different composition is done to have a better matrix material

2.1. *1ST PROCESS (Al 85%, Graphite 10%, MoS2 5%):* In this process we are used aluminum Pieces for aluminum melting. The melting process was done by using the furnace. The weight machine is used for measuring aluminum pieces. Before measure the aluminum piece we set the weight machine at 0 reading. The aluminum pieces are filled in the cup and measure percentage in weight machine. Before filling into the cup, we place the cup on the weight machine and we should set the reading as 0. Then we filled the aluminum pieces in the cup in the measurement of ½ kg of aluminum 85%. Then we measure graphite at 10% and MoS2 at 5%.Then we move to furnace and we set the furnace temperature at 950°C. Then we filled the measured aluminum pieces in the furnace in Crushable cup. The aluminum filled in 2 times of furnace. On First time we filled 50% aluminum in the Crushable cup then close the furnace door after 20 min the aluminum pieces are melted. Then we filled the balance aluminum pieces that is 35% in the Crushable cup of furnace then close the door, in this time we open the furnace door after 15 min. The total aluminum pieces are melted then we add the graphite 10% in the crushable cup in that time we mixed the graphite powder and melted aluminum by using mixing blade in furnace in this process going to 10 min. this process is called by STIR CASTING.After 10 min we open the furnace then add the MoS2 powder 5% in the crushable cup. Then we mix the aluminum, graphite, MoS2 by using mixing blade. The STIR CASTING process is done within 10 Min. Then furnace indicate alarm in 950°C the process was done. Then the crushable cup hold by using the metal grip holder. Then we pour the hot metal in die. Then the die was cooled 1 hour in outside. To take the aluminum rod the die should be braked. Then turning process done by lathe. The aluminum rod is fabricated. In this result the rod is full of black powder because the materials are not mixing properly[5].

Fig.1. Fabrication of Aluminum Graphite and MOS2 Composite Materials

2.2. *2nd PROCESS (Al 93%, Graphite 5%, MoS2 2%):*we filled the aluminum pieces in the glass in the estimation of aluminum 93%. At that point we measure graphite at 5% and MoS2 at 2%. At that point we move to heater and we set the heater temperature at 950°C. At that point we filled the deliberate aluminum pieces in the heater in Crushable glass [1]. The aluminum filled in multiple times of heater. On First time we filled 63% aluminum in the Crushable container at that point close the heater entryway after 20 min the aluminum pieces are dissolved. At that point we filled the equalization aluminum pieces that is 30% in the Crushable measure of heater at that point close the entryway, in this time we open the heater entryway after 15 min. The all out aluminum pieces are liquefied then we include the graphite 5% in the crushable container in that time we blended the graphite powder and softened aluminum by utilizing blending sharp edge in heater in this procedure going to 10 min. this procedure is called by STIR CASTING. After 10 min we open the heater at that point include the MoS2 powder 2% in the crushable glass. At that point we blend the aluminum, graphite, MoS2 by utilizing blending cutting edge. The STIR CASTING process is done inside 10 Min. At that point heater show alert in 950°C the procedure was finished. At that point the crushable glass hold by utilizing the metal grasp holder. At that point we pour the hot metal in bite the dust. At that point the kick the bucket was cooled 1 hour in outside. To take the aluminum pole the kick the bucket ought to be braked. At that point turning process done by machine. The aluminum bar is manufactured. In this outcome the pole is loaded with gaps that is porosity.

2.3. *3rd PROCESS (Al 96%, Graphite 2%, MoS2 2%):*we filled the aluminum pieces in the container in the estimation of aluminum 96%. At that point we measure graphite at 2% and MoS2 at 2%. At that point we move to heater and we set the heater temperature at 950°C. At that point we filled the deliberate aluminum pieces in the heater in Crushable glass. The aluminum filled in

IOP Conf. Series: Materials Science and Engineering **574** (2019) 012001 doi:10.1088/1757-899X/574/1/012001

multiple times of heater [2]. On First time we filled 66% aluminum in the Crushable container at that point close the heater entryway after 20 min the aluminum pieces are softened. At that point we filled the parity aluminum pieces that is 30% in the Crushable measure of heater at that point close the entryway, in this time we open the heater entryway after 15 min. The all out aluminum pieces are softened then we include the graphite 2% in the crushable container in that time we blended the graphite powder and dissolved aluminum by utilizing blending edge in heater in this procedure going to 10 min. this procedure is called by STIR CASTING. After 10 min we open the heater at that point include the MoS2 powder 2% in the crushable glass. At that point we blend the aluminum, graphite, MoS2 by utilizing blending cutting edge. The STIR CASTING process is done inside 10 Min. At that point heater show caution in 950°C the procedure was finished. At that point the crushable glass hold by utilizing the metal grasp holder. At that point we pour the hot metal in pass on. At that point the kick the bucket was cooled 1 hour in outside. To take the aluminum pole the kick the bucket ought to be braked. At that point turning process done by machine. The aluminum bar is created. In this outcome it conquer superior to past two procedure [4].

3. COMPOSITION STUDY

By using stir casting process these material in combined at efficient ratio as shown in the table. This process is done three times using different composition.

Table 1 (AL85%+C10%+MOS2 5%)

Calculating	Values
Density	2.7
Young's modulus	70
Poisson's ratio	0.35
Thermal conductivity	237
Tensile strength ultimate	75
Compressive strength ultimate	210
Bonded	5%mixing

Table 2 (AL93%+C5%+MOS2 2%)

Calculating	Values
Density	2.9
Young's modulus	68
Poisson's ratio	0.31
Thermal conductivity	234
Tensile strength ultimate	67
Compressive strength ultimate	215
Bonded	75%mixing

Table 3 (AL96%+C2%+MOS2 2%)

Calculating	Values
Density	3
Young's modulus	65
Poisson's ratio	0.32
Thermal conductivity	211

Tensile strength ultimate	60
Compressive strength ultimate	209
Bonded	85%mixing

4. TESTING CONDITION

- HARDNESS TEST
- BENDING TEST

HARDNESS TEST - The assurance of the Rockwell hardness of a material includes the utilization of a minor burden pursued by a noteworthy burden as shown in Fig 2. The minor burden sets up the zero position. The real burden is connected, at that point evacuated while as yet keeping up the minor burden. The profundity of entrance from the zero datum is estimated from a dial, on which a harder material gives a higher number. That is, the infiltration profundity and hardness are contrarily corresponding. The main favourable position of Rockwell hardness is its capacity to show hardness esteems legitimately, consequently deterring dreary estimations engaged with different hardness estimation systems. The condition for Rockwell Hardness is HR=N-d/s, where d is the profundity (from the zero burden point), and N and s are scale factors that rely upon the size of the test being utilized (see following segment). It is normally utilized in designing and metallurgy. Its business prominence emerges from its speed, unwavering quality, vigour, goals and little region of space [3].

Fig.2 HARDNESS TEST

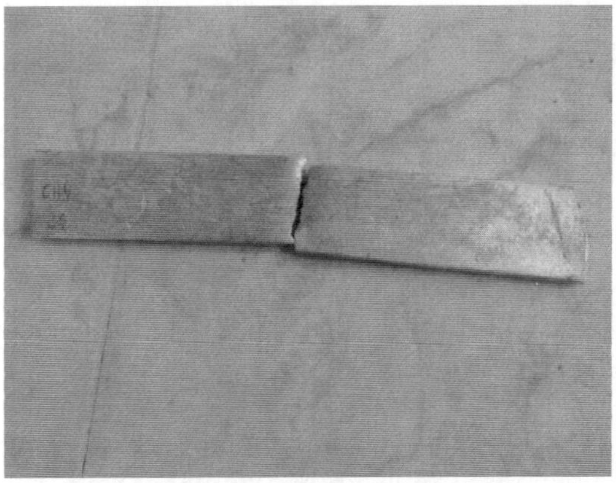

Fig.3 BENDING TEST

BENDING TEST-In applied mechanics, bending (also known as flexure) characterizes the behaviour of a slender structural element subjected to an external load applied perpendicularly to a longitudinal axis of the element as shown in Fig 3. A beam deforms and stresses develop inside it when a transverse load is applied on it. In the quasi-static case, the amount of bending deflection and the stresses that develop are assumed not to change over time. In a horizontal beam supported at the ends and loaded downwards in the middle, the material at the over-side of the beam is compressed while the material at the underside is stretched. There are two forms of internal stresses caused by lateral loads. They are shear stress and compressive

IOP Conf. Series: Materials Science and Engineering **574** (2019) 012001 doi:10.1088/1757-899X/574/1/012001

5. RESULT

5.1. Rockwell Hardness test

Specification: AL+C+MOS2
Test method: ASTM E18:2016
Rockwell Hardness Result = 63(HR15N)

5.2. Bending test

Test method: ASTM A 370:2017
Load at peak: 2.760 KN
Bend strength : 13.988 N/mm^2

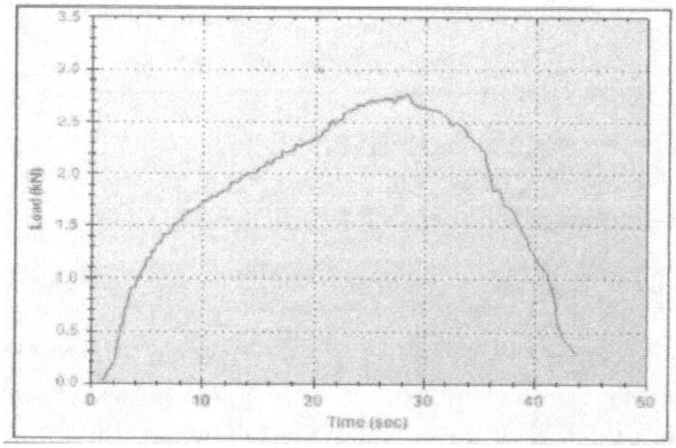

Fig.4 BENDING TEST RECORD

6. CONCLUSION

In this project we fabricated Aluminum rod in different composites (Aluminum, Graphite, and MoS2). The result of 1st stage is full of black powder on aluminum rod in the 2nd stage is full of holes on the aluminum rod and in the 3rd stage it better than previous two process. We are doing the better process for the good finished aluminum rod.

References

[1] Rohatgi, A., Soulami, A., Stephens, E. V., Davies, R. W., and Smith, M. T., 2014, "An investigation of enhanced formability in AA5182-O Al during high-rate free-forming at room-temperature: Quantification of deformation history," Journal of Materials Processing Technology

[2] Higashi, K., Mukai, T., Kaizu, K., Tsuchida, S., and Tanimura, S., 1991, "Strain rate dependence on mechanical properties in some commercial aluminum alloys," Le Journal de Physique IV

[3] Higashi, K., Mukai, T., Kaizu, K., Tsuchida, S., and Tanimura, S., 1991, "The microstructural evolution during deformation under several strain rates in a commercial 5182 aluminium alloy," Le Journal de Physique IV

[4] N. Chawla and Y.L. Shen, "Mechanical Behavior of Particle Reinforced Metal Matrix Composites," in Advanced Engineering Materials, 3(6), (2001), pp. 357–370

[5] Davis, J. R., 1993, Aluminium and Aluminium Alloys, ASM International

[6] N. Barekar, S. Tzamtzis, B.K Dhindaw, J. Patel, N. HariBabu, and Z. Fan, "Processing of Aluminium-Graphite Particulate Metal Matrix Composites by Advanced Shear Technology," in Journal of Materials Engineering and Performance, 18(9), (2009), pp. 1230–1240

IOP Conf. Series: Materials Science and Engineering **574** (2019) 012002 doi:10.1088/1757-899X/574/1/012002

Optimizing the Concrete Materials by Taguchi Optimization Method

Sabarish K.V[a], Akish Remo M[b], Pratheeba Paul[c]
[a]Vels Institute of Science, Technology & Advanced Studies, Chennai, India
[b]Student, Department of Civil Engineering, Vels University Chennai, India
[c]Hindustan Institute of Technology and Science, Chennai, India

ABSTRACT: This manuscript presents the outcome of an experimental investigation conceded out to optimize the mix proportions of the concrete by the Taguchi method of parameter design. The concrete mix design is done by the L9 orthogonal array that is 34. When a new material is used in the concrete all the mix variables are taken into account so that it may cost so much, it means more need for money, material, personnel and time. In order to reduce the number of mix combination in concrete, the Taguchi method is used. In this project the cement sand and aggregate are partially replaced by cement as fly ash, river sand as M-sand and the compression values are taken and the value is compared with the conventional concrete special concrete compression value is compared with the conventional concrete and S/N value is taken from the compression value and the comparison of the values are done. In order to attain good strength and to avoid cracks, sisal fiber is added in the concrete with a percentage of (1, 1.5, and 2).

1. INTRODUCTION

In this, the Taguchi method are used to optimize the combination of a selected parameters because it has so many numbers of combination that is a full factorial method and it is difficult to do project so that Taguchi method are used to optimize the number of combination to make the project cost-effective.

The concrete mix designs are done by the L9 orthogonal array that is 34. When a new material is used in the concrete all the mix variables are taken into account so that it may cost so much, it means more need for money and material, personnel, time. In order to reduce the number of mix combination in concrete, the Taguchi method is used In this the materials are partially replaced with cement to fly ash, Sand to m-sand, and coarse aggregate is replaced by the selected percentage and sisal fiber is added to avoid the cracks fiber is mixed with the concrete and cubes are casted.

The concrete cube should be properly cast and the compaction should be done properly so that no air bubbles will not be present inside the concrete cube. After the concrete cube are cast. The cubes are cured and they are tested for 7 days, 14 days and 28-day compression test. This test is done in a compression machine. The outcome of an experimental study carried out to optimize the mix size of the concrete by the Taguchi method. The compression values that are conventional concrete and special concrete compression value are compared with the conventional concrete so that we will get a compression graph and S/N value is taken from the compression mean value. The mean value and the S/N values are plotted in the graph so the graphs are compared and we can get a final value.

2. MATERIAL PROPERTIES

Cement

Ordinary Portland cement (7.5 Grade) was used in all test specimen. Properties are tested by referring IS 12269-1987. The specific gravity of cement was 3.13. Standard consistency of cement is 34%. The fly ash was used as a partial replacement of cement by different proportions.

Fly ash

Fly ash is one of the binding materials used in this project with cement and it is partially replaced with cement.

Fine Aggregates

The fine aggregate utilized was the normal sand free from pollutions and passing through 4.75 mm IS sieve. The specific gravity was found to be 2.65, and the fine aggregate is partially replaced with M-sand.

Coarse Aggregates

The locally available coarse aggregate passing through 20 mm IS sieve and retained on 10 mm IS sieve was used. The specific gravity of coarse aggregate was found to be 2.8.

Fibers

In this study sisal fiber is used for the project and these fibers are taken from the cactus plant leaf and its fibers are added with the percentage of 1, 1.5 and 2.

3. METHODOLOGY

Taguchi technique is a tool for optimizing the recital characteristic of a process. The aim of the research is to spot and design the method parameters that optimize the chosen quality characteristic and they are least to noise factors. In the current study, the goal of the job is to see the effects of the method parameters on the concert and the optimum blend of control factors that would increase the compressive strength of the concrete which is elected as the quality attribute.

Selection of control factors and their levels are made on the basis of some literature review. Four control factors are in use such as water/binder–ratio, fly ash, M-sand, and Sisal fibers are selected for the project. The elected four control factors are treating at three levels, as exposed in Table 1. The choice of three levels has been concluded because, due to the effect of these factors on the show characteristic may vary.

An orthogonal array is a fractional design which will balance property. By array design, the property of multiple variables on the concert characteristics can be predictable at the same time while minimize the quantity of the test. An L9 (3^4) orthogonal array is selected for the present examination S/N ratios are three types, i.e. the smaller is better, the higher is better, and the nominal is finest. The formula worn to design for analyzing S/N percentage, it is revealed below.

Smaller is better

It is chosen when the goal is to minimize the response. The S/N can be calculated for smaller the better

$$\frac{S}{N} ratio = -10\, log_{10}\, \frac{1}{n}\sum_{i=1}^{n} yi^2$$

Larger is better

It is chosen when the goal is to maximize the response. The S/N is calculated for larger the better.

$$\frac{S}{N} ratio = -10\, log_{10}\, \frac{1}{n}\sum_{i=1}^{n} \frac{1}{yi^2}$$

Nominal is better

It is chosen when the goal us to target the response and it is required to base the S/N on the standard deviation only. The S/N is calculated for smaller the better.

$$\frac{S}{N} ratio = -10\ log_{10} \frac{1}{n} \sum_{i=1}^{n} (y_i - y_0)^2$$

4. RESULT AND DISCUSSION

The universal testing machine is used to measure the compressive strength of the concrete cubes. The three compression readings are recorded for each combination as shown in Table 5. These parameters are taken from a designed experiment to analyze the mean function.

Table 1. Parameters and levels

Factors		**Level 1**	**Level 2**	**Level 3**
Water/ binder ratio	A	0.40	0.45	0.5
Fly ash (%)	B	20	25	30
M Sand (%)	C	30	35	40
Sisal Fiber (%)	D	1	1.5	2

Table 2.Full factorial orthogonal array for L_9

Experimental No.	**Factor A**	**Factor B**	**Factor C**	**Factor D**
1	1	1	1	1
2	1	1	2	3
3	1	1	3	2
4	1	1	2	1
5	1	1	3	3
6	1	1	1	2
7	1	1	1	3
8	1	1	2	2
9	1	1	3	1
10	1	2	1	2
11	1	2	1	3
12	1	2	1	1
13	1	2	2	2
14	1	2	2	3
15	1	2	3	3
16	1	2	3	1
17	1	2	3	2
18	1	2	2	1

IOP Conf. Series: Materials Science and Engineering **574** (2019) 012002 doi:10.1088/1757-899X/574/1/012002

19	1	3	1	1
20	1	3	2	3
21	1	3	3	2
22	1	3	2	1
23	1	3	3	3
24	1	3	1	2
25	1	3	1	3
26	1	3	3	1
27	1	3	2	2
28	2	1	2	3
29	2	1	1	3
30	2	1	3	2
31	2	1	2	2
32	2	1	1	2
33	2	1	2	1
34	2	1	1	1
35	2	1	3	3
36	2	1	3	1
37	2	2	3	1
38	2	2	2	3
39	2	2	1	3
40	2	2	3	2
41	2	2	1	2
42	2	2	2	2
43	2	2	2	1
44	2	2	1	1
45	2	2	3	3
46	2	3	1	1
47	2	3	2	3
48	2	3	1	3
49	2	3	3	2
50	2	3	2	2
51	2	3	1	2
52	2	3	3	3
53	2	3	3	1
54	2	3	2	1
55	3	1	3	1
56	3	1	3	2
57	3	1	1	3
58	3	1	1	2
59	3	1	2	1
60	3	1	1	1
61	3	1	2	2
62	3	1	3	3
63	3	1	2	3
64	3	2	3	1
65	3	2	1	3

66	3	2	1	2
67	3	2	2	1
68	3	2	3	2
69	3	2	2	3
70	3	2	1	1
71	3	2	2	2
72	3	2	3	3
73	3	3	2	1
74	3	3	3	1
75	3	3	3	2
76	3	3	1	3
77	3	3	3	3
78	3	3	2	2
79	3	3	1	2
80	3	3	2	3
81	3	3	1	1

Generally, there are three standard S/N equations are widely used to classify the objective function- larger the better, smaller the better or nominal the best.

In the present study compressive strength should be larger so that and our goal is also to maximize the strength. The standard S/N ratio is given below.

$$\frac{S}{N}ratio = -10\ log_{10}\ \frac{1}{n}\sum_{i=1}^{n}\frac{1}{yi^2}$$

Using the above formula for each of the nine experimental combinations and are reported in Table 6. The comparison between the compression value of conventional concrete and special concrete is plotted in the graph, as shown in figure 1. In the compression graph, the highest value is combination 4which has the highest value of nine combinations. The standardideals of S/N ratios of the four control factors at all of the levels are exposed in Figure 2, and from which the levels resultant to the highest S/N ratio values are chosen for all parameter instead of the optimum circumstance. Here, the optimum condition is the maximization of the compressive strength. In the Figure 2 the optimum levels are: A2 (water/binder ratio: 0.45), B2 (fly ash 25%), C2 (coarse sand: 35%) and D3 (stone dust: 2%) respectively.

Table 3.Standard orthogonal array $(L_9)3^4$

Experimental No.	Factor A	Factor B	Factor C	Factor D
1	1	1	1	1
2	1	2	2	2
3	1	3	3	3
4	2	1	2	3
5	2	2	3	1

6	2	3	1	2
7	3	1	3	2
8	3	2	1	3
9	3	3	2	1

Table 4.Taguchi optimization experimental design $(L_9)3^4$

Experimental No	A Water binder/ratio	B Fly ash (%)	C M sand (%)	D Sisal fiber (%)
1	0.40	20	30	1
2	0.40	25	35	1.5
3	0.40	30	40	2
4	**0.45**	**20**	**35**	**2**
5	0.45	25	40	1
6	0.45	30	30	1.5
7	0.50	20	40	1.5
8	0.50	25	30	2
9	0.50	30	35	1

The standard values of the compressive strength for every factor at the three levels have been designedand are plotted in Figure 3. The mean response analysis also indicate the similar optimum level of the parameters (A2, B2, C2, and D3) as is obtain in S/N ratio analysis. So that the strength which is attained is correct.

Table 5. Compression result and S/N ratio value

Experimental No.	COMPRESSIVE STRENGTH OF MATERIALS (KN)				S/N ratio (dB)
	Reading 1	Reading 2	Reading 3	Mean	
1	130	150	175	151.667	43.584
2	225	255	240	240.000	47.541
3	245	260	270	258.333	48.219
4	340	355	370	355.000	50.970
5	235	250	245	243.333	47.695
6	260	245	255	253.333	48.007
7	183	220	195	199.333	46.000
8	290	303	315	302.667	49.581
9	237	243	225	235.000	47.389

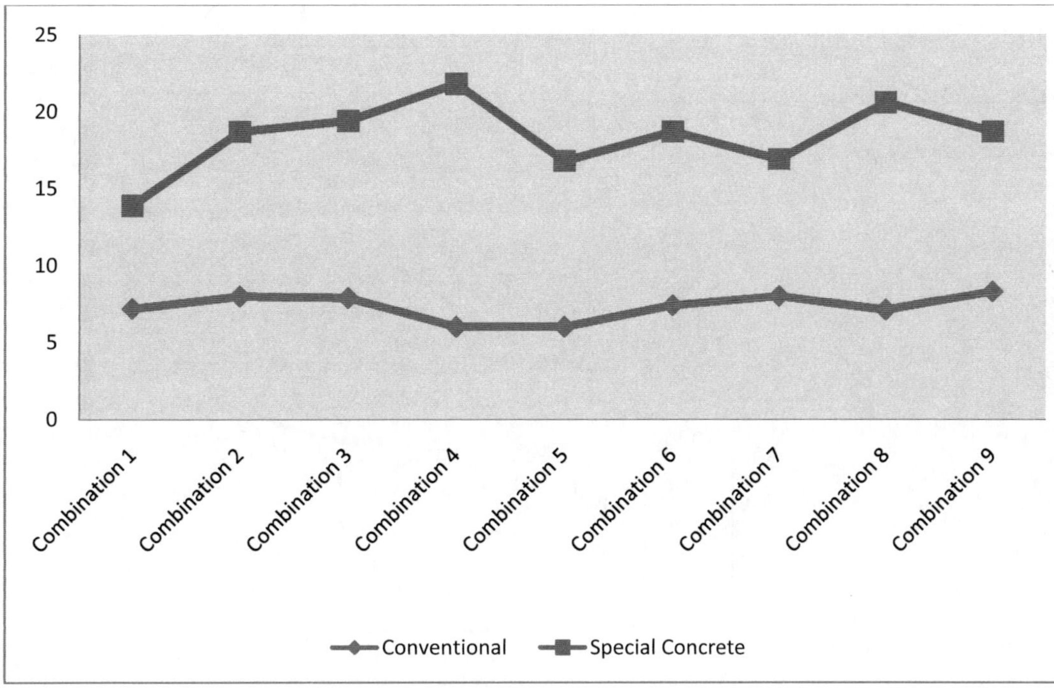

Fig 1.Compression graph of conventional vs special concrete

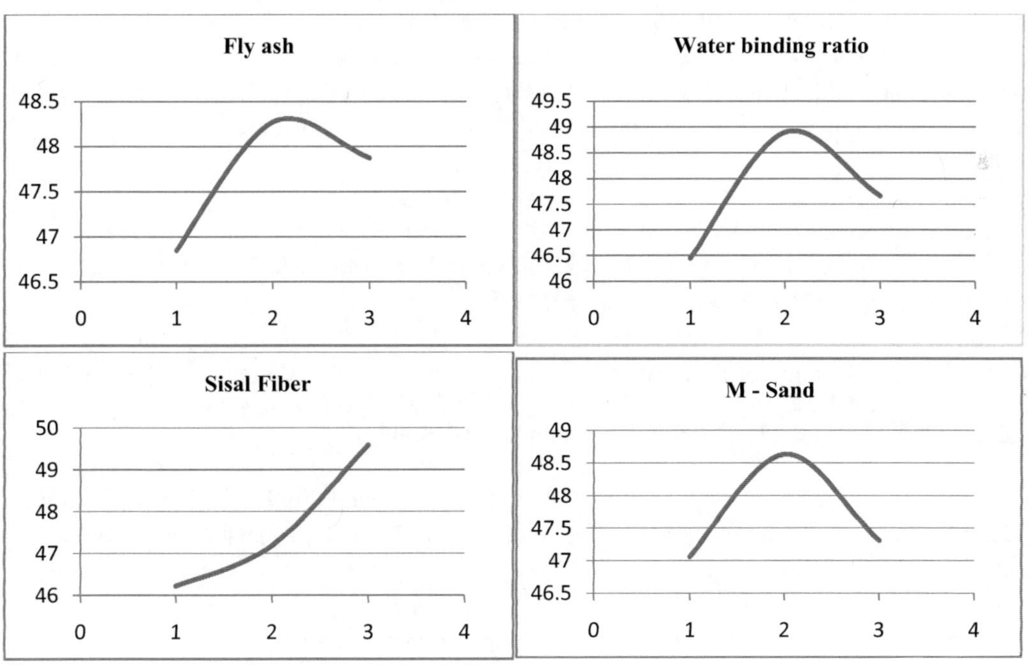

Fig 2.Process parameters on average S/N ratio

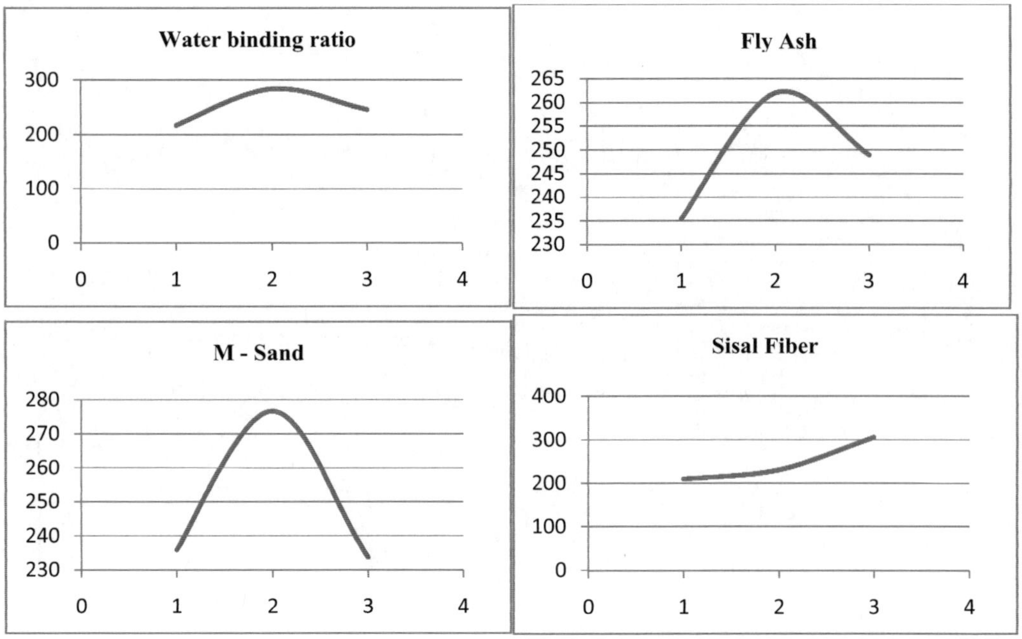

Fig 3.Mean response characteristic graph

CONCLUSION

The parameters of the mix design are optimized by Taguchi method for an L9 orthogonal array with four control factors and three levels. Selected parameters along with the levels are; water/ binding ratio (0.40, 0.45, 0.50); fly ash (20, 25, 30); M-sand (30, 35, 40); sisal fiber (1, 1.5, 2). In this orthogonal combination(2, 1, 2, 3) gives more strength when compared with other orthogonal combination. The total material cost of the project and the time conception is highly reduced.

REFERENCES

[1] Sabarish.k.v,Experimental Studies On Partial Replacement Of Cement With Fly Ash In Concrete Elements, International Journal of Civil Engineering & Technology (IJCIET), Volume 8, Issue 9, September 2017, pp. 293–298.

[2] Sabarish.k.vUtilization of Sisal Fiber In Portland Cement Concrete Elements, “International Journal of Civil Engineering & Technology (IJCIET), Volume 9, Issue on the 9th of September 2018, page 1682–1686- Scopus Indexed Journal.

[3] Sabarish.k.v, Experimental Investigation on Strength and Durability Properties of Sisal Fiber Reinforced Concrete, International Journal of Chemical Sciences, August & 2016, PP - 241-246.

[4] Chang PK, Hou WM. A study on the hydration properties of high performance slag concrete analysed by SRA. Cement and Concrete Research 2003; 33(2):183-189.

[5] Singh H, Kumar P. Tool wear optimization in turning operation by Taguchi method. Indian Journal of Engineering and Material Sciences 2004; 11(1):19-24.

[6] Sabarish, Experimental Investigation on Strength and Durability Properties of Sisal Fiber Reinforced Concrete SSRG International Journal of Civil Engineering – (ICCREST'17) - Special Issue- March 2017,PP-6569.

[7] Nuruddin, M.F., Bayuaji R., Mirha Foamed Concrete For Load Bearing Structures, in International Engineering Convention 2009 (Intec2009). 2009: Damascus Syria.

IOP Conf. Series: Materials Science and Engineering **574** (2019) 012003 doi:10.1088/1757-899X/574/1/012003

An Efficiency Study On Water Extraction From Air Using Thermophoresis Method

Dhanraj. S[a], Dr.Harish[b], R.Gopi[c] and Ganesh Babu Loganathan[d]

[a]Student, Department of Mechanical Engineering, Saveetha School of Engineering, Chennai, Tamil Nadu, India.

[b]Assistant Professor, Department of Mechanical Engineering, Saveetha School of Engineering, Chennai, Tamil Nadu, India.

[c]Assistant Professor, Department of Mechanical Engineering, Adhi College of Engineering and Technology, Kanchipuram, Tamil Nadu, India.

[d]Department of Mechatronics Engineering, Tishk International University, Erbil, KRG, Iraq.

Corresponding author: dhanrajsundaram.1995@gmail.com

Abstract : Only three-dimensional of the world's water is H2O, and common fraction of that's tucked away in frozen glaciers or otherwise out of stock for our use. As a result, some 1.1 billion individuals worldwide lack access to water, and a complete of two.7 billion realize water scarce for a minimum of one month of the year.An AWG extracts water from air.Water vapour within the air are often extracted by condensation - cooling the air below its saturation point, exposing the air to desiccants. At the same time applying thermophoresis method where the system is cooled to its max and surrounding atmospheric air is forcefully heated such that during there contact water droplets formation is increased and efficiency is increased which would be inversely proportional to the time taken. Even if applying cascade system of refrigeration, the efficiency of water generation from humid air is increased.

1. INTRODUCTION

This paper is taken into account as a concept to scale back the water scacrcity in several regions. Over 1.2 billion individuals lack access to scrub potable. The speed at that water may be created depends on ratio and close air temperature and size of the mechanical device. Region water generators become simpler as ratio and air temperature increase. The cost-effectiveness of Associate in Nursing AWG depends on the capability of the machine, native wetness and temperature conditions and also the value to power the unit. AWG is Associate in Nursing already existing device. The water extraction capability differs in keeping with the models. By applying thermophoresis technique wherever it had been found that the water extraction from air at its traditional weather temperature with constant wetness is accrued n times (n depends on condenser temperature,evaporator temperature and alternative weather factors) of water extraction from air by forcefully increasing traditional temperature of weather air temperature. The efficiency of water extraction from air is completed by thermophoresis take a look at and condensation method. The efficiency study is completed by testing at completely different temperature and scrutiny them. The test has been done using both thermophoresis method and non thermophoresis method and the graphs are compared. Using thermophoresis method of heat exchange- the hot air pass is more, the temperature inside is less,the water droplets are more so the rate of water production is more.The weight is less compared to oldsetup. Eneregy consumption is less and the cost is also low.

2. ARCHITECTURE

The design consists of a condenser,compressor,cooling and evaporating coils,filters, storage tank and outlet. In a cooling condensation sort atmospherical water generator, a mechanical device circulates refrigerant through a condenser and so an evaporator coil that cools the air close it.This lowers the air temperature to its temperature, causing water to condense.A controlled-speed fan pushes filtered air over the coil.The ensuing water is then passed into a holding tank with purification and filtration system to assist keep the water pure and cut back the chance expose by viruses and bacteria which can be collected from the close air on the evaporator coil by the compression water.The rate at that water will be made depends on ratio and close air temperature and size of the mechanical device.

Atmospheric water generators become more practical as ratio and air temperature increases. This means they're comparatively inefficient once placed within cool offices.The cost-effectiveness of an AWG depends on the capability of the machine, native wetness and temperature conditions and therefore the price to power the unit.

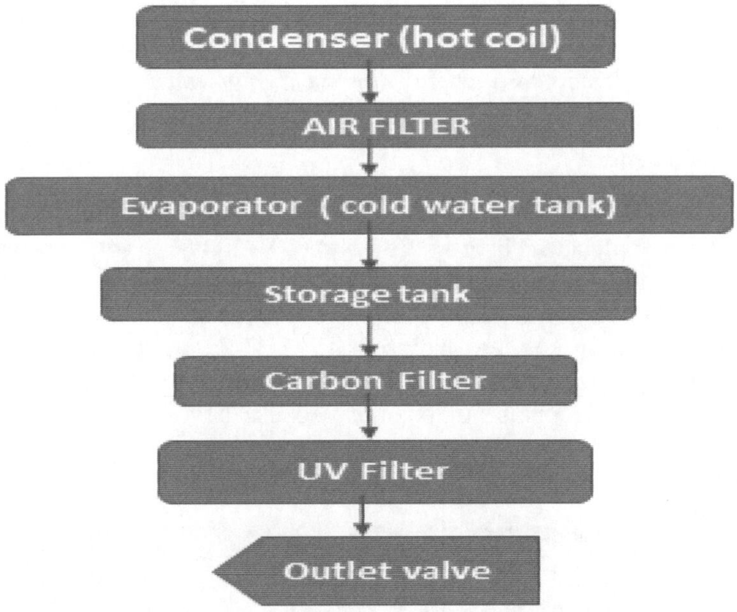

Fig 1. Block Diagram for Aqua Boy

CONDENSER

A condenser is a device to condense a substance from its air form to a liquid form by allowing it to cool. In thus doing, the heat is given up by the substance and transferred to the encompassing atmosphere.Condensers will be created per various styles, and are available in several sizes starting from rather tiny (hand-held) to terribly massive (industrial-scale units utilized in plant processes).For example, a icebox uses a condenser to induce eliminate heat extracted from the inside of the unit to the skin air. Condensers ar utilized in air conditioners, industrial chemical

IOP Conf. Series: Materials Science and Engineering **574** (2019) 012003 doi:10.1088/1757-899X/574/1/012003

processes like distillation, steam power plants and alternative heat-exchange systems.Use of cooling water or close air because the fluid is common in several condensers.

EVAPORATOR

An evaporator is a device which is used to turn the liquid form of a chemical substance like water into its gaseous-form/vapor. In this process the liquid is evaporated or vaporized into a gas form.

AIR FILTER

A particulate air filter is a device composed of fibrous or porous materials that removes solid particulates like dirt, pollen, mold, and bacterium from the air.Filters containing associate degree adsorbent or catalyst like charcoal (carbon) might also take away odors and vaporific pollutants like volatile organic compounds or gas. Air filters are employed in applications where air quality is vital, notably in building ventilation systems and in engines. Some buildings, in addition as craft and different human-made environments (e.g., satellites and area shuttles) use foam, folded paper, or spun covering material filter components. Another technique, air filters use fibers or components with a static charge, which magnetize dirt particles.The air intakes of burning engines and air compressors tend to use either paper, foam, or cotton filters.

CARBON FILTER

Carbon filter refers to a way of filtration that uses carbon to separate the impurities of water or air. Carbon filters use a process where the pollutants adhere to the carbon particles as the water or air passes through it. Carbon filters are useful for purifying water and purifying the air of contaminants, but they are also exceptionally useful in removing offending odors from and around indoor gardens.

UV FILTER

UV filters are mixtures that block or absorb ultraviolet (UV) light-weight.Two major applications of actinic radiation filters area unit in sunscreens, wherever they shield the skin from sunburn and different injury, and in photography, wherever they scale back the extent of ultraviolet light that strikes the recording medium. When exposed to UV light, these filters can undergo transformations into other chemicals that are less protective and possibly toxic. As a result, there are practical concerns about the choice of chemical to use beyond just which has the desired filtering properties.

Fig 2. Experimental Setup

Fig 3. Thermophorosis

RESULTS AND OBSERVATION

Table 1. PROPERTIES OF R-134a

DENSITY	0.00425 g/cm
BOILING POINT	-14.9^{0}F or -26.1^{0}F
AUTO IGNITION TEMPERATURE	1418^{0}F or 770^{0}F
OZONE DEPLETION LEVEL	0
SOLUBILITY	0.15%

T= Temperature
P= Pressure
h = Enthalpy
s = Entropy
x = Quality(0 = liquid,1 = vapour)
m = Mass flow rate
m = 3 kg/s

State 2 :
T = 50.9^{0}C
P = 1200kPa
h = 279.3kJ/ kg
s = 0.9301kJ/ kg.K

IOP Conf. Series: Materials Science and Engineering **574** (2019) 012003 doi:10.1088/1757-899X/574/1/012003

High Pressure, High Temperature and Superheated Vapour
State 3 :
T = 46.3⁰C
P = 1200kPa
h = 117.8kJ/ kg
s = 0.4245kJ/ kg.K

High Pressure, Medium Temperature and Saturated Liquid

Non Thermophoresis Test
The day of experiment room temp was noted as 28 degree Celsius with 73% relative humidity. The temperature of the cold body was noted as 11 degree Celsius and was constant during experiment. Under non thermophoresis method when the cold body was substituted in a room with temperature of 28 degree Celsius and 73% humidity and the time taken for water droplet formation for 5 readings.

Table 2. Non Thermophoresis Test

TEMPERATURE	HUMIDITY	TIME TAKEN (min/s)
28⁰C	73%	1.38
28⁰C	73%	1.42
28⁰C	73%	1.35
28⁰C	73%	1.52
28⁰C	73%	1.52

Average of 5 readings = 1.438

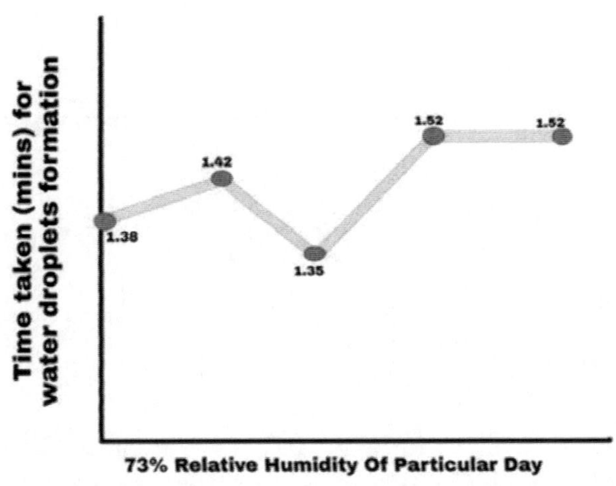

Fig 4. Non Thermophoresis Method
(experiment done in a room temperature of 28^0C and 73% of relative humidity on a cold body with the temperature of 11^0C)

THERMOPHORESIS TEST

Under thermophoresis method the cold body was substituted under the room with temp of 31 deg Celsius (the air inside the room was forcefully passed through the condenser coil to increase its temperature) with relative humidity of 73% and time taken for five readings.

Table 3. Thermophoresis Test

TEMPERATURE	HUMIDITY	TIME TAKEN (min/s)
32^{0} C	73%	1.22
32^{0} C	73%	1.25
32^{0} C	73%	1.31
32^{0} C	73%	1.25
32^{0} C	73%	1.23

Average of 5 readings = 1.252

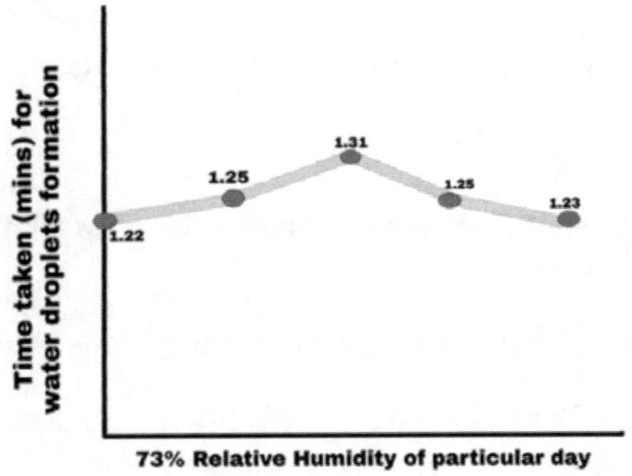

Fig 5. Thermophoresis Method
(experiment done in a room temperature of 31^{0}C and 73% of relative humidity on a cold body with the temperature of 11^{0}C)

Calculating the change in percentage of time taken for water droplet formation on cold by non thermophoresis method and thermophoresis method.

Avg. non thermophoresis method – Avg. thermophoresis method / avg.time taken *100

1.438-1.252/1.252*100 = 14.85%

CONCLUSION :
This paper elaborates the design and construction of the atmospheric water generator and the study of its efficiency using thermophoresis method. The readings are taken using thermophoresis method and non thermophoresis and compared at different temperatures. The extraction of water from air is done by thermophoresis test and condensation process.

REFERENCES

[1] Environmental Assessment of Air to Water Machines. International Journal of Life Cycle Assessment, 18:1149-1157.

[2] Patents; Draw water from air, live what quantity water you drink and be kind to the fish you catch, New York Times. July 2, 2001

[3] "Solar-Powered Device Pulls Water Out of skinny (And Pretty Dry) Air", Retrieved 13 April 2017.

[4] This widget Makes Gallons of potable Out of Air Time INC, April 24, 2014

[5] Latest Willie Nelson venture: Water from Air. Atlanta Journal Constitution.

[6] Water Extracted from the Air for Disaster Relief. National Public Radio; by Nell Greenfieldboyce; October 19, 2006

[7] Innovation Awards: Ahead of the Pack. Wall Street Journal. October 30, 2007.

[8] Drinking Water From Air Humidity. ScienceDaily (June 8, 2009)

[9] "2016 Toyota Mirai Fuel-Cell Sedan". Retrieved 28 August 2016.

[10] "New rooftop solar hydropanels harvest drinking water and energy at the same time". Retrieved 2017-11-30.

[11] "Rain fed solar powered water purification systems". Retrieved 21 October 2017.

[12] Categories: Drinking waterWater supplyWater and the environment, Duhr S, Braun D (December 2006).

[13] "Why molecules move along a temperature gradient". Proc. Natl. Acad. Sci. U.S.A. 103 (52): 19678–19682. Bibcode:2006PNAS..10319678D. doi:10.1073/pnas.0603873103. PMC 1750914. PMID 17164337.

IOP Conf. Series: Materials Science and Engineering **574** (2019) 012004 doi:10.1088/1757-899X/574/1/012004

A Study on Key Factors Influencing Bid Decision Model for Construction Projects

Kiran Kumar.J, Vimal Raj
Assistant Professor, Saveetha School of Engineering, Saveetha Institute of Medical and Technical Sciences, Chennai

Corresponding author: kirankumar.jagdish@gmail.com

Abstract. Competitive bidding serves as a platform for public/private contractors to execute the works specified by the client. Deciding to bid or not to bid for that job proves to be a major hindrance between the bidder and the work. Smart contractors tend to bid for a work, based on facts studied through a pre-bid survey rather being heuristic in taking their decisions. This paper identifies the potential factors influencing the contractors' bid decisions for different construction projects through a structured questionnaire survey. The factors were ranked based on their importance and utilized as a base for developing a bid decision model using DEA, a non-parametric approach. These factors govern the framework of bid decision model being developed. DEA creates a favorability score from the top and least ranked factors for each and every bid opportunity with a cut-off value as a reference. The public and private bidders utilize the cut-off value as a pivotal point of making bid decisions. This study serves as a tool for decision making process whenever a bid opportunity is created. This paper also recommends the contractors to utilize the bid model, to not only execute the work successfully, but also to improve the self-credibility and effectiveness among the global market.

Keywords: Bidding, DEA, Decision, Credibility, Heuristic, Construction

1. Introduction

Out of all sectors in India, construction industry accounts for a 19% of GDP growth, making it a major contributor for the development of the nation's economy. Most of the public works are tendered through a bidding process and the person with the least bid value (in terms of cost) procures the project. Deciding to bid or not to bid for the specified work is quite vital in the initial phases of the project as it challenges the contractor's identity among the global market. Decision making poses a major obstacle between the contractor and the bid which has a greater impact over the successful completion of the project and profit obtained. In general contractors are heuristic in nature while bidding for a project, deciding based on their experience and familiarity of the specific work. Smart contractors on the other hand consider the internal and external factors influencing the bid decision process rather being instinctive on their decisions.

In today's competitive market, the contractor needs to have a perfect bidding strategy to obtain and execute the work efficiently. In order to develop a suitable strategy which ensures higher rate of efficiency, the contractor must be aware of all the internal and external factors governing the bidding process. This strategy analyses the factors governing the bidding process, which acts as a base for developing a bid decision model. This bid decision model not only helps in contractor's decision making process but also proves to be a reference tool for future projects to be obtained through bidding. The Data Envelopment Analysis (DEA) proves to be an effective tool for developing a decision model as it increases the success rate in winning a bid. With this development of the bid decision model, the contractors will be able to analyze the current bid scenario and make a decision to bid or not for a particular project.

2. Literature study

Bagies and Fortune (2009) identified that out of all the respondents, 95% agree considering the factors influencing the bid decisions, not only to be critical but also improves the performance of the company. Large number of internal and external factors governing the decisions of a contractor makes the bidding process quite a complex one. The study recommended an implementation of optimum mark-up model in facing the challenges at the time of bidding. Nyoman Yudha Astana et.al (2015) developed a bidding strategy by surveying the importance of the factors governing the bidding process and validated the data using Confirmatory Factor Analysis (CFA) and calculated the loading factor by adopting Structural Equation modelling of partial least squares. Economic situation, competition, client characteristics and company experience were found to be the most influential factors from the study. Shash (1993) reported that many of the top contractors in United Kingdom decided to bid for a project and their mark-up value based on a subjective assessment conducted through a questionnaire survey. The weightage of all the factors were calculated using importance index and the results showed that, experience with the specific work and strength in the industry enhances the contractors' confidence in winning the project. Ahmad and Minkarah (1987) identified the top factors affecting the contractors in United Kingdom by conducting a questionnaire survey. The data collected were assessed subjectively to identify the potential factors affecting the bid decisions of the contractors. D.K.Chua and D.Li (2000) developed a bid reasoning model using Analytic Hierarchy Process (AHP) with the data collected through a questionnaire, surveyed over 153 contractors. The findings of the study suggested that the bidding failure occurs due to over estimation of cost to do the work and formulated the hierarchies based on competition, risk, need for work and company's position in bidding. Oyeyipo et.al (2016) collected the factors through a questionnaire and determined the significant differences of the means of the factors which have a greater impact over the bidders' decision. Availability of capital, financial capability of client, degree of hazard was found to be the vital factors affecting the bid decisions of indigenous contractors. S.Ping.Ho and Y.Liu (2004) showcased the dynamic nature between the claims and opportunistic bidding by developing a framework using Claims Decision Model. This model analyses the decisions and the construction claoms rationally and systematically. R.Sammoura (2010) developed a simulation model for bid decisions using Crystal Ball decision engineering software. This software considers the number of bidders and size of the project as the vital governing factors while developing the model. El-Mashaleh (2013) reported the weightages and the deviation aspect of the factors governing the bidding process along the

Jordan area. This study analysed the concordance of respondents using Cronbach alpha relaiability test and Kendall's ranking consensus. The study reported the statistical difference among the private and public bidders using Analysis of Variance tool. Egemen and Mohamed (2007) reported the completeness of bid documents, risks due to inflation as the important factors considered by the top contracting firms over the Northern Cyprus region in facing the bid process.

3. Methodology

Formulating a decision framework necessitates the study of the potential factors influencing the entire bidding process. An effective development of a decision framework requires a methodology suiting the conditions relating to the current bidding scenario under study. This study identifies the key factors through a structured questionnaire survey conducted among the top contractors in India. The methodology of this study is shown in the flowchart figure.1

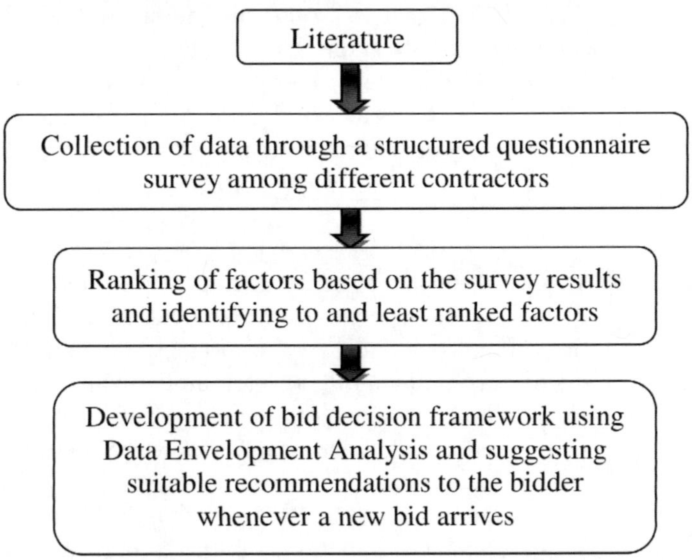

Figure.1. Methodology for development of bid decision model

4. Potential factors influencing the bidding decision

Developing a framework model necessitated the assimilation of key factors influencing the decision making process of a contractor. The top ranked factors (tf) and the least ranked factors (lf) are shown in the table 1 and table 2 respectively.

Table.1 Top Ranked Factors

Description	Top ranked factors
tf 1	Project (total bid value)
tf 2	Amount of possible upcoming profitable projects in future
tf 3	Current financial condition of the company
tf 4	Having qualified material suppliers
tf 5	Project type
tf 6	Experience and familiarity with the specific work

Table.2 Least Ranked Factors

Description	Least ranked factors
lf 1	Possessing enough qualified technical staff to do the job
lf 2	Desire of qualified contractors to bid and win the project
lf 3	Uncertainty related to construction site
lf 4	Need for continuity in employment of key staff personnel
lf 5	Current workload in bid preparation
lf 6	Technological difficulty of the project beyond the capability of the firm

The questionnaire was formed with the potential factors governing the bidding scenario, as shown above their level of importance was determined with a score of 0 to 6 ranging from least importance to high importance and. The results of this survey forms the basis of developing a bid decision model which is explained with a real time project obtained through competitive bidding. A sample of questionnaire survey and the rate at which the responses were recorded are as follows,

How far does the current financial situation of the company affect the contractors' bidding decision?

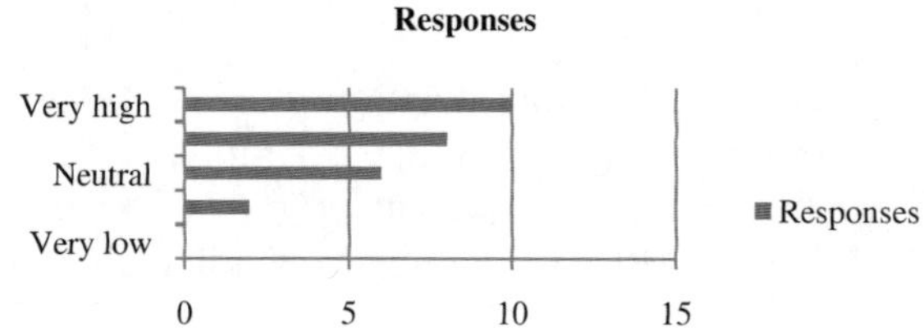

Figure.2. Number of responses from the survey

5. Data envelopment analysis

Data envelopment analysis is non-parametric linear programming model developed to measure the productive efficiency of the Decision Making Units (DMU). The DMU are the core study under the DEA approach. Compared to other areas of study, the utilization of this analysis is in its nascent stages in construction industry. DEA analyses the decision making units and this framework is governed by inputs and outputs as the former are minimized and (or) the latter are maximized. The DEA is executed in the following stages.

- Top ranked and least ranked factors are identified from the survey conducted.
- The ranked factors are rated on a scale of 0 to 10 ranging from least to highly important based on the current bidding scenario.
- The favourability score is calculated from the inputs and outputs given and compared with the cut-off value which consequently indicates the contractor to bid or not to bid for that particular works.

An envelopment surface is developed with the DEA approach, which paves way in determining decision making units (DMU). The DMU within the determined envelope are considered to be favorable bidding conditions. The framework of the decision making unit is shown in the Figure.3.

Figure.3. Input / Output Decision making unit

The efficiency of the decision making unit are calculated as follows

$$\frac{weightedsumofoutputs}{weightedsumofinputs} = \frac{\sum_{i=1}^{n} u(i)y(iq)}{\sum_{j=1}^{m} v(j)x(ij)}$$

where, v (j) = 1, 2,....m are the input weights

u (i) = 1, 2…n are the output weights

This efficiency value is compared with a cut-off value which acts as a reference value computed from the previous bids stored in the contractor's database. A real time project obtained through bidding process after the Data Envelopment Analysis is discussed. The bid under consideration is a provision of a sewage treatment unit at TBM colony and provision of RCC tank of 90000 liters capacity, under corporate welfare plan. A total of five contractors submitted their bids and the contractor with the least bid procured the work. The details and the specifications of the proposed bid are shown in table 3.

Table.3. Proposed bid details for Overhead Tank

PROPOSED WORK	TOTAL BID VALUE (IN RUPEES)	BIDDER DETAILS	SELECTED BIDDER
Provision of RCC overhead tank of 90000 liters and sewage treatment unit	94,16,917	M.RAMESH C.SARAVANAN A.SADAGOPAN N.SWAMI C.JAGADEESAN	M.RAMESH

The contractor obtained the bid by submitting the least value (in terms of cost) for the proposed work using Data Envelopment Analysis. The scores of the top ranked and the least ranked factors provided by the winning bidder and other competitors are shown in the table 4 as follows

Table.4Proposed bid details for Overhead Tank

Bidder details	Top ranked factors					Least ranked factors				
	tf1	tf2	tf3	tf4	tf5	lf1	lf2	lf3	lf4	lf5
M.Ramesh (B1)	7	4	4	9	6	5	2	4	8	3
C.Saravanan (B2)	5	3	2	5	1	3	2	4	5	4
A.Sadagopan (B3)	7	5	5	4	3	4	5	3	9	5
N.Swami (B4)	4	2	5	4	3	1	7	6	6	8
C.Jagadeesan (B5)	8	4	1	5	5	5	4	4	7	4

The favorability scores are calculated with the equation mentioned above and compared with the cut-off value for each bidder, as shown in Table 5.

Table.5 Favorability score of all bidders

Bidder details	Favorability scores
B1	1.363
B2	0.889
B3	0.923
B4	0.642
B5	0.958

6. Results

With the scores from all the bidders, the favorability score is calculated and from the above tabulation, the bidder B1 with a value of 1.363, obtained the bid as the efficiency was greater than the cut-off value (1.0). This DEA approach proposes the contractor to bid by providing high scores for top ranked factors and low scores for least ranked factors so that the probability of winning a bid is high in order to obtain an effective model.

7. Conclusion

- This study reports the potential factors influencing the bid decisions of the contractors which act as a base for the development of the decision model. Total bid value, project type, availability of qualified material suppliers, current financial position are the most influential factors compared to uncertainty related to the site, technological difficulty and current workload in bid preparation.

- A public work procured through competitive bidding is considered to understand the way Data Envelopment Analysis works. The DEA approach aids the contractors in analysing the current situation whenever a new bid comes. The favourability score is calculated and compared with the pre-determined cut-off value to determine whether to bid or not to bid for the work under study.
- This DEA approach proves to be developmental force to the industry as it finds its applications over vast areas, regardless of the size, location of operation and the number factors governing the bidding process.

8. Acknowledgement

I hereby acknowledge with deep sense of gratitude the valuable guidance, encouragement and suggestions given by my guide, company personnel and faculty of Civil Engineering Department of Saveetha School of Engineering, SIMATS who has been a constant source of inspiration throughout this project work.

References

[1] Ahmad, I., and Minkarah, I., (1988), Questionnaire survey on bidding in construction, Journal of management in engineering ASCE, 4(3), 229-243.

[2] Chua, D. K. H. and Li. D. (2000), Key factors in bid reasoning model, Journal of Construction Engineering and Management, September, 349-357.

[3] El-Mashaleh, M. (2013), Empirical framework for making bid/no-bid decision, Journal of Management in Engineering, 29(3), 200-205.

[4] Irtishad Ahmad et.al (2013), Questionnaire Survey on Bidding in construction, ASCE, 229-243.

[5] Mohamed S. El-mashaleh, et.al (2014), Understanding Key bidding factors considered by top Jordanian Contractors, Vol.8, 455-464.

[6] Nyoman Yudha Astana, et.al (2015), Factors affecting Bidding Strategy in Construction, 9(5), 211-215.

[7] O.O. Oyeyipo, et.al (2016), Factors affecting contractors' bidding decisions for construction projects in Nigeria, 1-21.

[8] Bageis, A. S., & Fortune, C. (2009), Factors affecting the bid/no bid decision in the Saudi Arabian construction contractors, Construction management and economics, 27(1), 53-71. DOI: 10.1080/01446190802596220.

[9] M. Egemen and A. Mohamed, A framework for contractors to reach strategically correct bid/no bid and mark-up size decision, Building and Environment, Vol.42, 1373-1385, March 2007.

[10] Shash. A, (1993), Factors considered in tendering decisions by top UK contractors, Construction Management and Economics, 11(2), 111-118.

IOP Conf. Series: Materials Science and Engineering **574** (2019) 012005 doi:10.1088/1757-899X/574/1/012005

Synthesis of AgNPs Using Flowers Exhibiting Different Photoresponse

NK Udaya Prakash[a], N Sripriya[b], V Aishwarya[b], S Preethy[b], S Benetta[b] and S Bhuvaneswari[c]
[a]Department of Biotechnology, School of Life Sciences,
Vels Institute of Science, technology and advanced Studies, Chennai, India
[b]Research and Development, MARINA LABS, Chennai, India
[c]Department of Botany, Bharathi Women's College, Chennai, India

Corresponding author: nkudayaprakash@gmail.com

Abstract. Biological synthesis or green synthesis of nanoparticles has proven to be cost effective, rapid and safe. Flowers exhibit different colours due to wavelength selective absorption and light scattering. In this study, four differently coloured flowers, i.e. *Bauhinia purpurea* – Fabaceae, *Couroupita guianensis* – Lecithydaceae, *Gerbera jamesonii* – Asteraceae and *Gomphrena globosa* – Amaranthaceae were studied for their potency to synthesize silver nanoparticles. The UV-Vis spectrophotometric analysis of the aqueous extracts of *B. purpurea, C. guianensis, G. jamesonii* and *G. globosa* flowers recorded their respective peaks at 562, 498, 524 and 548 nm in response to their pigments. The synthesized silver particles were characterized through UV-Vis spectrophotometry, scanning electron microscopy, energy dispersive X-ray spectroscopy, particle size analyzer and fourier transform infrared spectroscopy. The resonance peak in the UV-Vis spectrum was recorded at 430, 420, 445 and 435 nm respectively, for the AgNPs synthesized using *B. purpurea, C. guianensis, G. jamesonii* and *G. globosa*. SEM analysis revealed the size range of 107-498 nm for *B. purpurea*, 15-75 nm for *C. guianensis*, 17-51 nm for *G. jamesonii* and 20-91 nm for *G. globosa*. The presence of silver was confirmed by EDS. The silver particles synthesized using the extracts of *B. purpurea, C. guianensis, G. jamesonii* and *G. globosa* showed the average size as 431.3, 71.51, 30.11 and 33.93 nm as assessed through PSA. FTIR studies confirmed the involvement of O-H, N-H, C-C=C, N=O, $-NO_2$ and C-O stretches. In the present study, the silver particles synthesized by *C. guianensis, G. jamesonii* and *G. globosa* were in the nanoscale, whereas those synthesized by *B. purpurea* were in microscale. The synthesis of AgNPs is attributed to the pigments or the metabolites present in the floral extracts. Further studies on the potential of flowers exhibiting different photoresponse in the synthesis of nanoparticles are recommended.

1. INTRODUCTION

The emergence of nanotechnology in several fields, such as catalysis, electronics, drug delivery, tissue engineering and as sensors is primarily due to the distinctive properties of the metallic nanoparticles (NPs), such as their high surface to volume ratio, surface plasmon resonance and other quantum effects [1]. The synthesis of NPs is performed by physical, chemical and biological methods. Biological synthesis or green synthesis of NPs has proven to be cost effective, rapid and safe [2]. Biological sources like bacteria [3], actinomycetes [4], fungi [5], algae [6] and plants [7] are highly exploited for the synthesis of NPs.

Rapid synthesis of NPs by plants is mainly attributed to the phytoconstituents, as reducing, capping and stabilizing agents [8]. Hence, studies on the usage of different plant parts are conducted to synthesize NPs [9]. Among the plant parts, the corolla of flower is found to exhibit different colours as photoresponse to light. The selective absorption of wavelength and light scattering by the floral components contributes to the floral colour [10]. The present study focuses on synthesis of silver nanoparticles (AgNPs) with four flowers exhibiting different colours as their photoresponse. The following flowers i.e. *Bauhinia purpurea* (Fabaceae), *Couroupita guianensis* (Lecithydaceae), *Gerbera jamesonii* (Asteraceae) and *Gomphrena globosa* (Amaranthaceae) are used in this study. Although many flowers such as *Nyctanthes arbortristis* [11], *Calotropis procera* [12], *Lonicera japonica* [13], *Rhododendron dauricum* [14], *Nelumbo nucifera* [15], *Saraca indica* [16], *Ipomoea indica* [17], *Achillea biebersteinii* [18], *Chrysanthemum indicum* [19], *Tagetes erecta* [20], *Plumeria alba* [21] and *Stenolobium stans* [22] have already been reported for the synthesis of NPs, the present study gives prime importance on the floral color in synthesis of AgNPs.

2. MATERIALS AND METHODS

2.1. Preparation of plant extract

The flowers of *B. purpurea*, *C. guianensis*, *G. jamesonii* and *G. globosa* (Fig. 1a-d) were collected from the flower market in Koyambedu, Chennai, India. Aqueous extracts of the flowers were prepared by boiling the corolla with distilled water at a ratio of 1:10 (w/v). The boiled solution was filtered and the filtrate was used for one step synthesis of AgNPs.

2.2. Synthesis of silver particles

Fifty ml of the aqueous extracts of the corolla was added to 500 ml of 1 mM silver nitrate ($AgNO_3$) solution. The solution was incubated in dark for 1 h and centrifuged at 3000 rpm for 10 min, to eliminate the unreacted compounds. The centrifuged supernatant was further washed with methanol and dried.

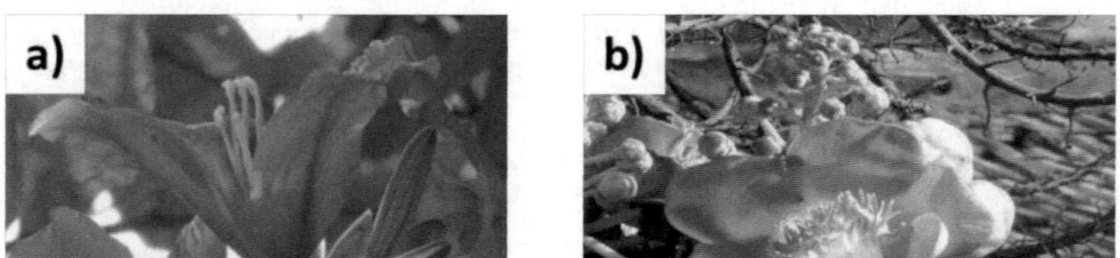

Fig. 1. Flowers of a) *B. purpurea* b) *C. guianensis* c) *G. jamesonii* and d) *G. globosa*

2.3. Characterization of silver particles

The synthesis of silver particles was ascertained by the surface plasmon resonance using UV-Vis spectrophotometer (Cyberlab, USA). The shape and size of the silver particles were observed through high resolution scanning electron microscopy (Quanta 200 FEG). The presence of silver in the particles synthesized was confirmed by EDS (Bruker Corp, USA). The average size of the particles was determined by particle size analyzer (Malvern Zetasizer Nano series). The functional groups involved in the synthesis of silver particles were identified using FTIR analysis (ALPHA-T, Bruker Corp, USA) in the wavenumber range of 4000 to 400 cm^{-1}.

3. Results

UV-Visible spectrophotometric analysis suggested that the floral extracts of *B. purpurea, C. guianensis, G. jamesonii* and *G. globosa* recorded their peaks at 562, 498, 524 and 548 nm respectively, in response to their floral pigments. Reduction of Ag^{+} ions to nanoscale is primarily visualized by the color change to reddish brown (Fig. 2a-d), which is due to the excitation of the surface plasmon. The resonance peak in the UV-Vis spectrum was recorded at 430, 420, 445 and 435 nm respectively, for the AgNPs synthesized using *B. purpurea, C. guianensis, G. jamesonii* and *G. globosa* (Fig. 3).

Fig 2. AgNPs synthesized using a) *B. purpurea* b) *C. guianensis* c) *G. jamesonii* and d) *G. globosa*

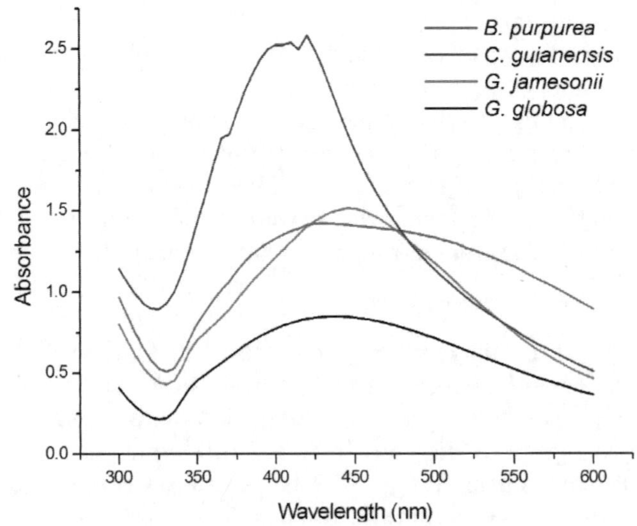

Fig 3. UV-Vis spectra of AgNPs synthesized using floral extracts of *B. purpurea, C. guianensis, G. jamesonii* and *G. globosa*

SEM analysis (Fig. 4a-d) revealed the silver particles to be of spherical to irregular shape with the size range of 107-498 nm for *B. purpurea*, 15-75 nm for *C. guianensis*, 17-51 nm for *G. jamesonii* and 20-91 nm for *G. globosa*. Analysis by EDS confirmed the presence of silver (Fig. 5a-d). Analysis of AgNPs through particle size analyzer revealed the average size of 431.3, 71.51, 30.11 and 33.93 nm for *B. purpurea, C. guianensis, G. jamesonii* and *G. globosa* respectively (Fig. 6a-d).

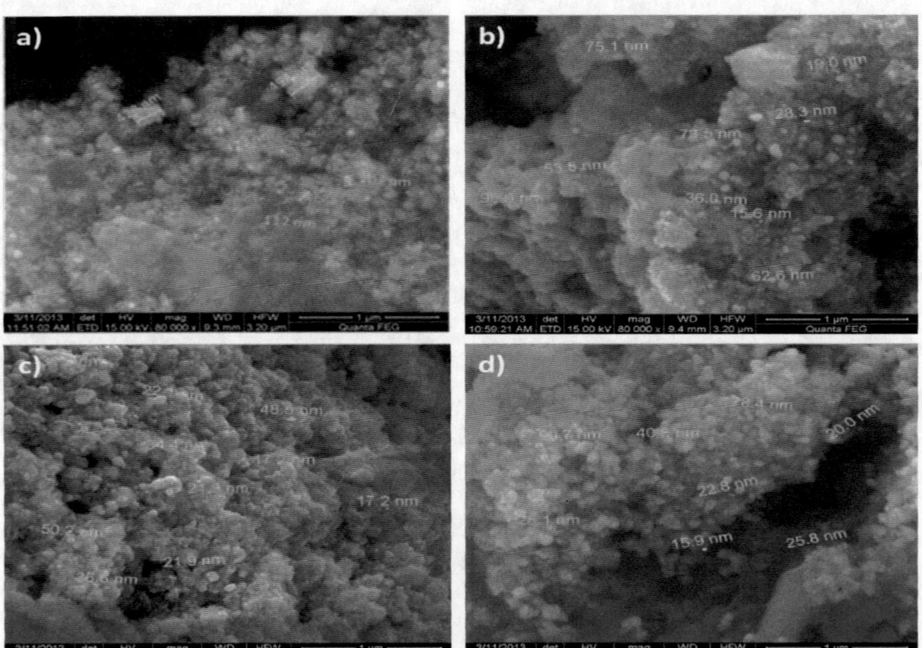

Fig 4. SEM images of AgNPs synthesized by a) *B. purpurea*, b) *C. guianensis*, c) *G. jamesonii* and d) *G. globosa*

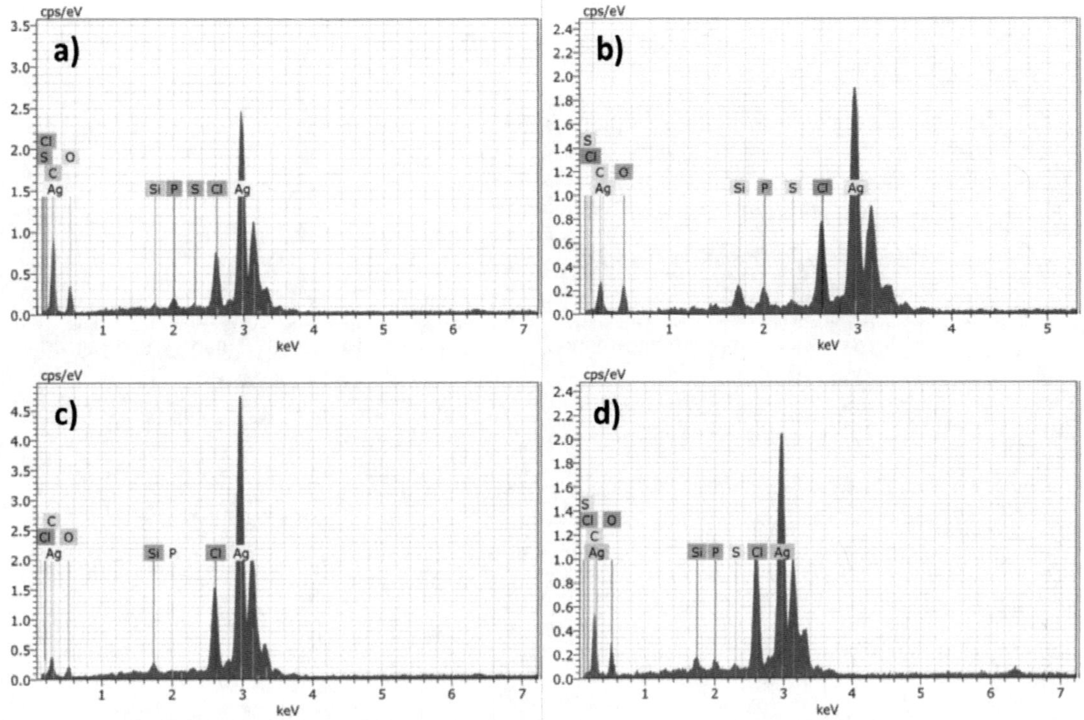

Fig 5. EDS analysis of AgNPs synthesized by a) *B. purpurea*, b) *C. guianensis*, c) *G. jamesonii* and d) *G. globosa*

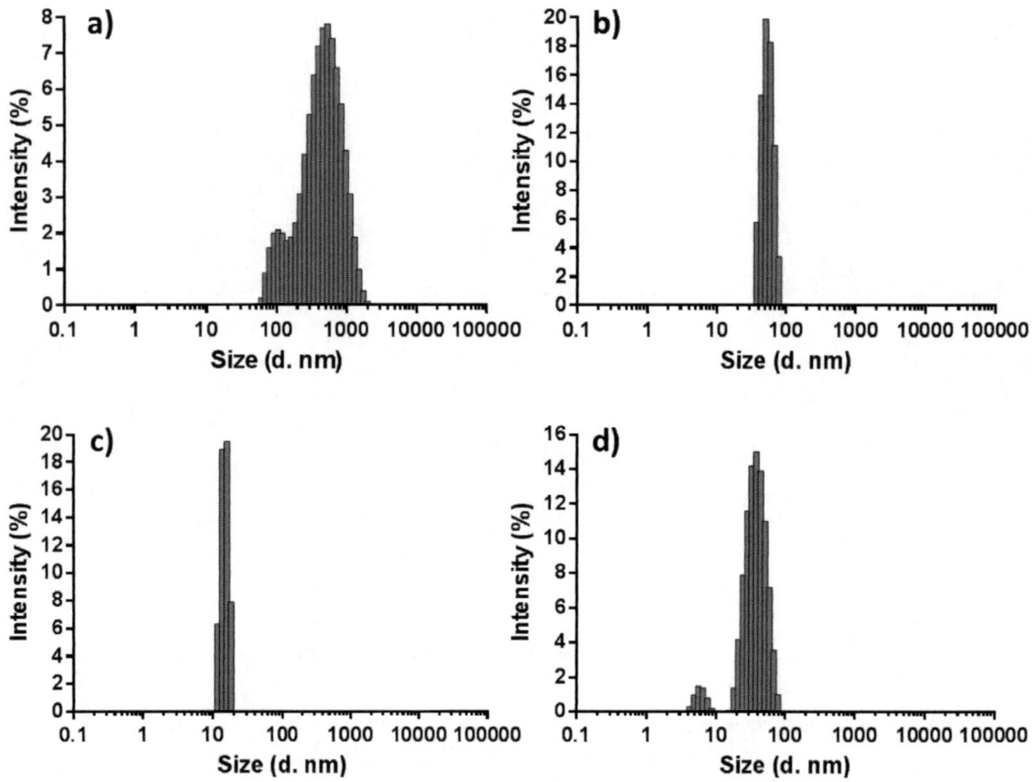

Fig 6. Particle size analysis of AgNPs synthesized by a) *B. purpurea* b) *C. guianensis* c) *G. jamesonii* and d) *G. globosa*

FTIR analysis (Fig. 7) recorded peaks at different wavenumbers for the AgNPs synthesized. The absorption bands at 3817-3513, 3482-3389, 1400 and 1367 cm^{-1} recorded for the AgNPs synthesized from *B. purpurea* corresponded to O-H of phenol/alcohol; N-H of secondary amine; O-H of carboxylic acids; N=O of nitro group and NO_2 of aliphatic group respectively. Similar peaks were recorded in the AgNPs from *G. globosa.* The peaks at 3975-3188, 2381, 1400 and 1367 cm^{-1} revealed the presence of O-H stretch of alcohol or phenol and carboxylic group and N-H stretch of amide groups in AgNPs synthesized from *C. guianensis*. In AgNPs synthesized from *G. jamesonii*, peaks were recorded at 3692-3409, 2687, 1400 and 1367 cm^{-1}, which are associated with O-H of phenol/alcohol; N-H of secondary amine; O-H of carboxylic acids; N=O of nitro group and NO_2 of aliphatic groups.

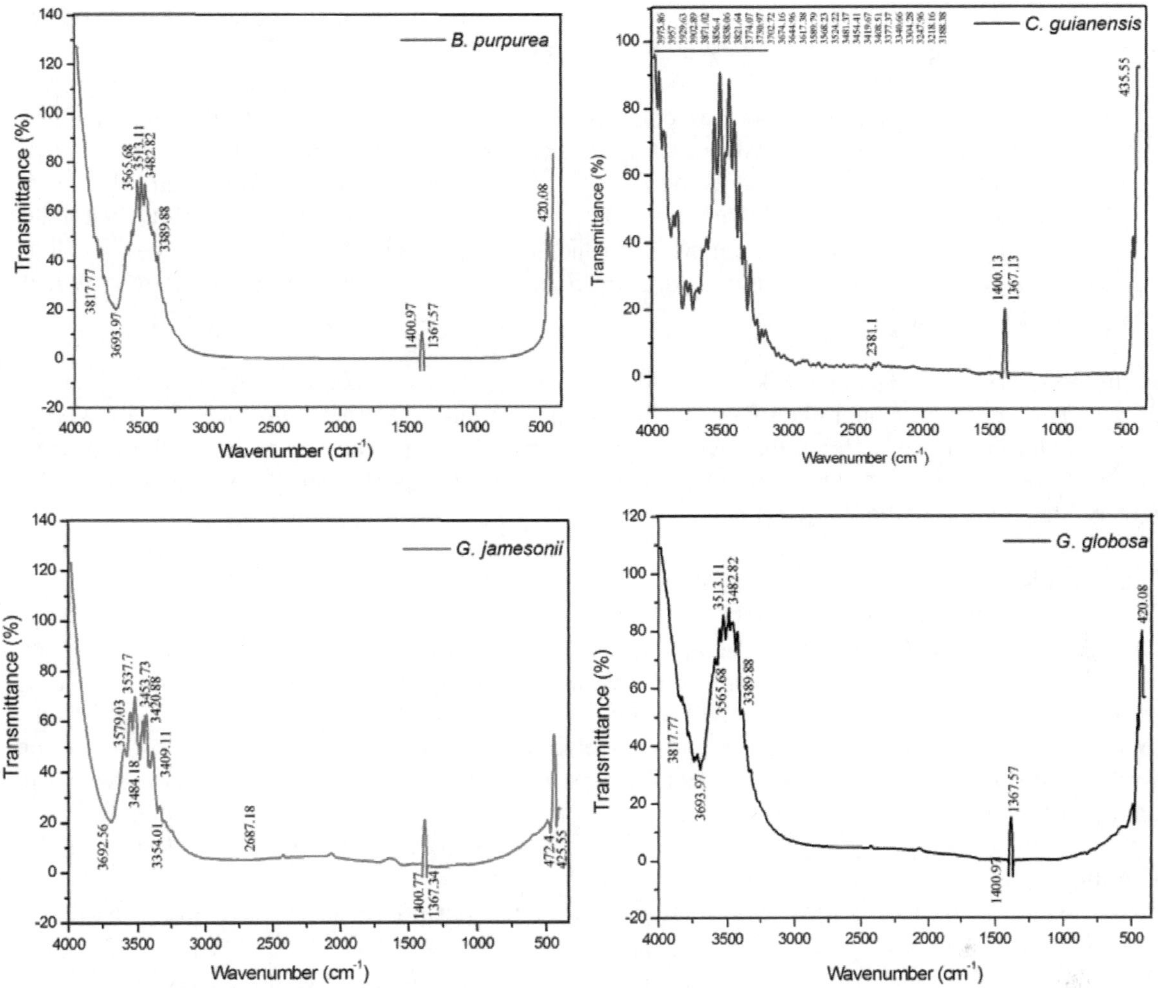

Fig. 7. FTIR spectra of AgNPs synthesized by the flowers of *B. purpurea*, *C. guianensis*, *G. jamesonii* and *G. globosa*

4. Discussion

The study demonstrated the synthesis of silver particles by the aqueous extracts of *B. purpurea, C. guianensis, G. jamesonii* and *G. globosa* flowers, exhibiting different photoresponse. The surface plasmon resonance recorded for the AgNPs synthesized using *B. purpurea, C. guianensis, G. jamesonii* and *G. globosa* was 430, 420, 445 and 435 nm respectively. According to Stamplecoskie and Scaiano [23], AgNPs record surface plasmon resonance at 420-450 nm. It is interesting to note that the size of AgNPs synthesized using different flowers belonging to different families differed drastically. Nanoparticles are defined as the particles exhibiting their size below 100 nm on any dimension [24]. In this study, the size of silver particles synthesized by *C. guianensis, G. jamesonii* and *G. globosa* were less than 100 nm, and hence in the nanoscale. However, the particles synthesized by *B. purpurea* showed their size in the range of 107-498 nm, i.e., in microscale.

The variation in colors of flowers is majorly due to their pigments present. It is known that the shades of yellow and orange color in flowers are due to the presence of carotenoids, anthoxanthins or anthochlors, or combinations of them. Red, violet and blue are due to the presence of anthocyanins and their modifications [25]. Plant based pigments are well known reducing agents and they play an important role in the synthesis of NPs [26].

Synthesis of AgNPs by the extracts of plant parts is due to their phytoconstituents. Tannins, flavonoids and terpenoids were recorded in *B. purpurea* flowers. In addition to the above compounds, saponins were recorded in the flowers of *C. guianensis* and *G. jamesonii*. However, in *G. globosa*, only saponins were recorded [27]. Plant polyphenols have the ability to donate electrons, which reflects their reducing power of metals [28]. The present study clearly shows that *B. purpurea* possesses less reducing nature and hence, less reduction in size of AgNPs when compared to other extracts.

In the present study, FTIR analysis revealed the presence of alcohol/phenol, carboxylic and amide groups in *C. guianensis* and phenol or alcohol group, secondary amine, nitro and aliphatic group in *B. purpurea.* This ascertains the involvement of phenolics, amine and amides as the reducing agent of Ag^{+} ions to Ag^{0} [29]. Thus, the synthesis of AgNPs is attributed to the pigments or the metabolites present in the floral extracts.

5. Conclusion

The present study reports the green synthesis of AgNPs using flowers exhibiting different photoresponse, viz., *B. purpurea, C. guianensis, G. jamesonii* and *G. globosa*. The silver particles synthesized by *C. guianensis, G. jamesonii* and *G. globosa* were in the nanoscale, whereas those synthesized by *B. purpurea* were in microscale. The differences in the particle size might be due to the functional groups of the pigments or metabolites present in the floral extracts. Further studies on the potential of flowers exhibiting different photoresponse in the synthesis of nanoparticles are recommended.

Funding
This research was supported by internal fund of R & D, Marina Labs, Chennai (Grant Number: ML-2016-RD002).

References

[1] Vilela D, González M C and Escarpa A 2012 Sensing colorimetric approaches based on gold and silver nanoparticles aggregation: Chemical creativity behind the assay. A review *Anal. Chim. Acta.*, 751, p 24-43.

[2] Zheng B, Kong T, Jing X, Odoom-Wubah T, Li X, Sun D, Lu F, Zheng Y, Huang J and Li Q 2013 Plant-mediated synthesis of platinum nanoparticles and its bioreductive mechanism *J. Colloid Interface Sci.*, 396, p 138-145.

[3] Pugazhenthiran N, Anandan S, Kathiravan G, Udaya Prakash N K, Crawford S and Ashokkumar M 2009 Microbial synthesis of silver nanoparticles by *Bacillus* sp. *J. Nanoparticle Res.*, 11, p 1811-1815.

[4] Udaya Prakash N K, Bhuvaneswari S, Prabha S B, Kavitha K, Sandhya K V, Sathyabhuvaneshwari P and Bharathiraja B 2014 Green synthesis of silver nanoparticles using airborne actinomycetes *Int. J. ChemTech Res.,* 6, p 4123-4127.

[5] Durán N, Marcato P D, Alves O L, De Souza G I H and Esposito E 2005 Mechanistic aspects of biosynthesis of silver nanoparticles by several *Fusarium oxysporum* strains *J. Nanobiotechnology,* 3, p 8-14.

[6] Castro, Laura, María Luisa Blázquez, Jesus Angel Muñoz, Felisa González A B 2013 Biological synthesis of metallic nanoparticles using algae *IET Nanobiotechnology*, 7, p 109-116.

[7] Vijayaraghavan K, Nalini S P K, Prakash N U and Madhankumar D 2012 One step green synthesis of silver nano/microparticles using extracts of *Trachyspermum ammi* and *Papaver somniferum Colloids Surfaces B Biointerfaces*, 94, p 114-117.

[8] Vijayaraghavan K, Nalini S P K, Prakash N U and Madhankumar D 2012 Biomimetic synthesis of silver nanoparticles by aqueous extract of *Syzygium aromaticum Mater. Lett.*, 75, p 33-35.

[9] Ganaie S U, Abbasi T, Anuradha J and Abbasi S A 2014 Biomimetic synthesis of silver nanoparticles using the amphibious weed *Ipomoea* and their application in pollution control *J. King Saud Univ. - Sci.*, 26, p 222-229.

[10] Stavenga D G and van der Kooi C J 2016 Coloration of the Chilean Bellflower, *Nolana paradoxa*, interpreted with a scattering and absorbing layer stack model *Planta*, 243, p 171-181.

[11] Das R K, Gogoi N and Bora U 2011 Green synthesis of gold nanoparticles using *Nyctanthes arbortristis* flower extract *Bioprocess Biosyst. Eng.*, 34, p 615-619.

[12] Babu S A and Prabu H G 2011 Synthesis of AgNPs using the extract of *Calotropis procera* flower at room temperature *Mater. Lett.,* 65, p 1675-1677.

[13] Nagajyothi P C, Lee S E, An M and Lee K D 2012 Green synthesis of silver and gold nanoparticles using *Lonicera japonica* flower extract *Bull. Korean Chem. Soc.*, 33, p 2609-2612.

[14] Mittal A K, Kaler A and Banerjee U C 2013 Free radical scavenging and antioxidant activity of silver nanoparticles synthesized from flower extract of *Rhododendron dauricum Nano Biomed. Eng.,* 4, p 118-124.

[15] Arokiyaraj S, Nyayiru Kannaian U P, Elakkya V, Kamala T, Bhuvaneswari S and Dinesh K V. 2013 Green synthesis of silver nanoparticles using aqueous floral extract of *Nelumbo Nucifera Mater. Sci. Forum,* 756, p 106-111.

[16] Vidhu V K and Philip D 2014 Spectroscopic, microscopic and catalytic properties of silver nanoparticles synthesized using *Saraca indica* flower *Spectrochim. Acta - Part A Mol. Biomol. Spectrosc.,* 117, p 102-108.

[17] Pavani K V., K G, Banerjee A and Suresh S 2013 Phyto-synthesis of silver nanoparticles using extracts of *Ipomoea indica* flowers *Am. J. Nanomater.* 1, p 5-8.

[18] Baharara J, Namvar F, Ramezani T, Hosseini N and Mohamad R 2014 Green synthesis of silver nanoparticles using *Achillea biebersteinii* flower extract and its anti-angiogenic properties in the rat aortic ring model *Molecules,* 19, p 4624-4634.

[19] Arokiyaraj S, Arasu M V, Vincent S, Prakash N U, Choi S H, Oh Y K, Choi K C and Kim K H 2014 Rapid green synthesis of silver nanoparticles from *Chrysanthemum indicum* L. and its antibacterial and cytotoxic effects: An *in vitro* study *Int. J. Nanomedicine,* 9, p 379-388.

[20] Padalia H, Moteriya P and Chanda S 2015 Green synthesis of silver nanoparticles from marigold flower and its synergistic antimicrobial potential *Arab. J. Chem.*, 8, p 732-741.

[21] Mata R, Reddy Nakkala J and Rani Sadras S 2015 Catalytic and biological activities of green silver nanoparticles synthesized from *Plumeria alba* (frangipani) flower extract *Mater. Sci. Eng. C,* 51, p 216-225.

[22] Udaya Prakash N K, Bhuvaneswari S, Sai Nandhini R, Abdul Azeez N, Al-Arfaj A A and Munusamy M A 2015 Floral synthesis of silver nanoparticles using *Stenolobium stans* L *Asian J. Chem.,* 27, p 4089-4091.

[23] Stamplecoskie K G and Scaiano J C 2010 Light emitting diode irradiation can control the morphology and optical properties of silver nanoparticles *J. Am. Chem. Soc.,* 132, p 1825-1827.

[24] Laurent S, Forge D, Port M, Roch A, Robic C, Vander Elst L and Muller R N 2008, Magnetic iron oxide nanoparticles: synthesis, stabilization, vectorization, physicochemical characterizations, and biological applications *Chem. Rev.,* 108, p 2064-2110.

[25] Geissman T A 1941 Flower coloration *J. Chem. Educ.,* 18, p 108-110.

[26] Egorova E M and Revina A A 2000 Synthesis of metallic nanoparticles in reverse micelles in the presence of quercetin *Colloids Surfaces A Physicochem. Eng. Asp.,* 168, p 87-96.

[27] Bhuvaneswari S, Deepa S, Sripriya N, Prameela L and Udaya Prakash N K 2014 Antioxidant activity and phytochemistry of various flowers from Tamil Nadu, India *Int. J. Res. Pharm. Sci.,* 5, p 40-45.

[28] Papuc C, Goran G V, Predescu C N, Nicorescu V and Stefan G 2017 Plant Polyphenols as Antioxidant and antibacterial agents for shelf-life extension of meat and meat products: Classification, structures, sources, and action mechanisms *Compr. Rev. Food Sci. Food Saf.,* 16, p 1243-1268.

[29] Shahaby O El, Zayat M El, Salih E, Sherbiny I M El and Reicha F M 2013 Evaluation of antimicrobial activity of water infusion plant-mediated silver nanoparticles *J. Nanomed. Nanotechnol.,* 4, p 178.

Mechanical behavior of nanoparticulate TiO_2 reinforced magnesium matrix composite

Ganesh Radhakrishnan, Anirudh P V, and Anujan Kumar S
Department of Mechanical Engineering, Velammal Engineering College, Chennai – 600 066.

Corresponding author: ganeshrk1982@gmail.com

Abstract. In this study, nano-sized titanium dioxide particulate reinforced pure magnesium matrix composite has been fabricated by powder metallurgy technique for five different weight fractions of reinforcement such as 4 wt%, 8 wt%, 12 wt%, 16 wt%, and 20 wt% respectively. Samples of these compositions are iso-statically cold compacted into cylindrical specimens of diameter and length each 15 mm respectively, followed by sintering and hot extrusion with an extrusion ratio of 1.56:1. SEM micrographs of blended composite powders before compaction confirmed the uniform distribution of reinforcement throughout the matrix without any agglomeration. Microstructural analysis of extruded specimens revealed the refined matrix grains and grain boundary. The reinforcement may interrupt the matrix grain growth on grain boundaries during extrusion. Targeting for biomedical applications like bone implants, kneecap surgery etc., the mechanical behavior of the composite such as microhardness, and compressive strength have been studied. It was observed from the study that the mechanical properties have been significantly improved by adding reinforcement compared to pure magnesium and better mechanical behavior have been noticed for composite with higher contribution of reinforcement.

Keywords: magnesium matrix composite, nanoparticulate, titanium dioxide, biomedical, bone implants, mechanical behavior

1. Introduction

During the past few years, the researchers focused majorly on magnesium due to its attractive low density. Even though it has low density and high strength it suffers from certain barriers such as low strength, poor ductility, poor wear, and corrosive nature which hindered it from being used in aerospace, automobile industries and much more fields in its pure form. Currently, researchers are working on adding reinforcements to magnesium metal matrix to improve its mechanical properties. Recent reviews highlighted that there are about 60 different types of components, from instrument panels to engine components, in which magnesium alloys such as cast alloys-AZ63, AZ81, AZ91 etc., wrought alloys-AZ31, AZ61, AZ80 etc., Elektron, Magnox, Magnuminium, Mag-Thor, Birmabright, Magnalium and ceramic reinforcements such as SiC, Al_2O_3, Si, Ti, Zr are added to magnesium are being developed for use. The use of magnesium alloys in automobile parts is predicted to increase at an average rate of 15% per year. Magnesium alloys are used in aircrafts, missiles, automobile wheels, housings, ladders, cell phones, power tools, textile machinery, seat frames etc. The next major field focusing on the usage of magnesium alloys is the biomedical field. Biomedical field has been using materials such as 316L SS, CoCrMo, Ti Ti6Al4V in all implants division ranging from cardiovascular to otorhinology. Implant alloys are fabricated using methods like investment casting, powder metallurgy, thermo-mechanical process(TMP), superplastic deformation (SPD), equal channel angular pressing (ECAP), accumulative roll-bonding (ARB), high-pressure torsion (HPT) etc. Magnesium can be used for bone implant since the modulus of the metal and bone is similar (40-45GPa) and hence secondary surgery for removal of the implant will not be necessary. But since pure magnesium is poor in certain mechanical properties, titanium dioxide is added to magnesium as reinforcement in powder metallurgy technique and the properties are studied.

Similarly, many other interesting recent works such as nano particulate of Zinc oxide and Zirconium oxide reinforcement in sisal fiber composite [1], Al_2O_3 reinforcement in coconut bunch fiber composite [2] and boron carbide reinforcement in aluminum alloy [3] have published. The following are the studies on material currently used in biomedical applications. Jian Fang Li et al had worked on 316L SS by layering the alloy of Zr and ZrO_2 in plasma surface alloying apparatus. The cross-section microscopy, surface, phase structure were analyzed and the surface roughness, wear were also measured. It is found that ZrO_2 layers tremendously enhance wear resistance, improve adhesion and spreading of osteoblast cells [4]. Sheida Shri et al had studied on the growth of tantalum multi-layer thin films on CoCrMo alloy done by radio frequency magnetron spattering system. The crystallography, hardness, Young's modulus, film thickness, hardness was studied. The result was that the substrate temperature and bias voltage influence alpha and beta-Ta formation [5]. M.M. Machado Lopez et al observed the property enhancement of 45s5 bioglass coating on Ti6Al4V alloy by a colloidal electrophoretic deposition process. The study included X-ray diffraction, chemical composition analysis, and coating evaluation. The result was that the corrosion resistance was increased and ensured good protection from ionic attack [6].

Several studies in magnesium reinforced with various particulates are discussed as follows. Manoj Gupta et al analyzed the Al and Mg-based nanocomposites synthesized using powder metallurgy and extruded for characterization studies. Density, microstructure characterization, X-ray diffraction analysis, tensile test, microhardness measurement was studied. It was observed that the hybrid reduced energy and time, hardness and work of fracture improved, and the processing time was considerably reduced [7]. Junko Umeda et al had studied on titanium particulate reinforced magnesium composite materials done by spark plasma sintering followed by hot extrusion. The optical microstructure, compression strength, tensile strength and X-ray diffraction were studied. The composite material resulted in increased compression strength but there was no improvement in tensile strength [8]. Ganesh Kumar Meenashisundaram et al had worked on titanium (IV) oxide nanoparticulate reinforced magnesium composite material done by disintegrated melt deposition technique followed by hot extrusion. The material characterizations studied were microhardness, compression, tensile, thermal expansion and fracture behavior. The composites resulted in increased hardness and fracture strain was increasing with increase in the addition of titanium (IV) oxide [9]. Cheng-Jie Li et al has worked on Mg-3Zn-0.2Ca-0.5Y alloy using extrusion of billet at different temperature after heat treatment. The compression strength, microstructural study, and tensile strength have been studied. It resulted in increased grain size with increase in extrusion temperature [10]. ShahrouzZamaniKhalajabadi et al have worked on magnesium titania-hydroxyapatite nanocomposites coated by magnesium oxide flakes and silicon by cold pressing with coating after heat treatment in argon atmosphere of cold pressed billets. The electrochemical, immersion, vitro biological and compression test were studied. The result was improved anticorrosion ability after surface modification [11]. AlirezaVahid et al had worked on magnesium reinforced with tantalum particles using sintering of compacted composite powder. Density, X-ray diffraction, microstructure study and thermal expansion behavior has been studied. The results of the above composites are increase in microhardness and compressive strength [12]. Dongfang Zhang et al has worked on hafnium coated magnesium alloy using sintering in tube furnace of hafnium coated magnesium. Crystal structure, electrochemical and corrosion test has been studied. The results are improved corrosion resistance with a change of microstructure and production of surface oxidation [13].

IOP Conf. Series: Materials Science and Engineering **574** (2019) 012006 doi:10.1088/1757-899X/574/1/012006

2. Experimental Procedure

2.1 Materials

Magnesium powder supplied by S D Fine-Chem Limited with mean particle size 177μm and >99.995% purity is used as the base material. Titanium Dioxide particles of mean size less than 45 μm and purity >99.5% supplied by Merck Life Science Private Limited is chosen as reinforcement.

2.2 Procedure

2.2.1 Particle Size Conversion

Titanium Dioxide particle with average size of 45 μm was converted to an average particle size of 100 nm using planetary mill which included a jar with 16 metal balls (ball to powder ratio is 5:1) for size conversion process. This process took 17 hours at 250 rpm in ball mill.

2.2.2 Elemental Mixture of Mg-TiO_2 Powders

Titanium Dioxide reinforcement was added to magnesium metal in five different weight fractions such as 4 wt%, 8 wt%, 12 wt%, 16 wt%, and 20 wt% respectively. This process of blending was carried out in the planetary mill at 200 rpm for 10 minutes for each composition.

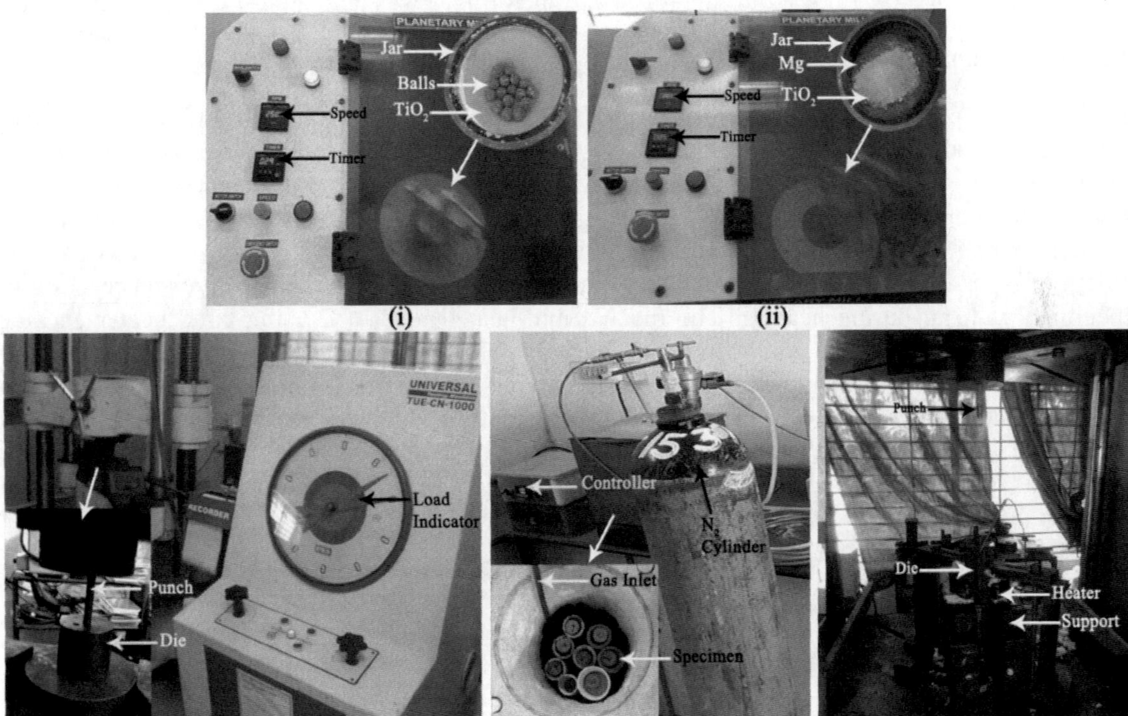

Fig.1 Sequence of operations – i) Particle size conversion ii) Elemental mixture iii) Compaction iv) Sintering v) Hot extrusion

2.2.3 Cold Compaction

The blended composite mixture was compacted at room temperature. The amount of powder weighed for each sample was 4.5 grams which produced a cylindrical billet of diameter and length of each 15 mm. This process was carried out in a Universal Testing Machine where the compaction die filled with weighed powders is fixed and a load of 50kN was applied using a punch. This load was held steady for 5 minutes to ensure that the load applied completely to produce a solid specimen.

2.2.4 Sintering

Sintering is a thermal treatment of fine-grained material at a temperature below the melting point of the base metal, for increasing its grain size and strength by bonding the particles together. The sintering process was done in two stages. In stage 1, the compacted specimens were sintered at 500°C for 4 hours in a muffle furnace at nitrogen atmosphere to prevent reaction of the specimen with air. A tray was prepared, and specimens were arranged in the tray in an individual crucible and placed inside the furnace. The gas was passed through a ceramic tube at 0.5 to 1 bar pressure into the chamber. Eight samples were heat treated at a time and took approximately 6 hours for one complete sintering process. The stage 2 sintering process is similar to stage 1, but the only difference is the heating temperature and the time of sintering were reduced. Sintering was done at 400°C for 1 hour at nitrogen atmosphere. The second stage sintering was done to refine grain and grain boundary. It took 1 hour to reach 400°C and the furnace was held at 400°C for 1 hour.

2.2.5 Hot Extrusion

The extrusion process is done to improve the bonding between matrix and reinforcement, refine matrix grain and also to refine grain boundary. An extrusion die had to be fabricated that supports the dimensions of the specimen. The extrusion temperature range for magnesium is 250°Cto 450°C. The reduction percentage was 20% in diameter i.e. 15mm diameter was reduced to 12mm through extrusion. And hence the extrusion ratio is 1.56:1. The taper angle of the die is 14° and length of taper is 10mm. The specimen and die were pre-heated separately to prevent reaction of specimen with air during heating. The die was pre-heated to 200°C using band heater and the specimen was pre-heated to 350°C in muffle furnace. The pre-heating of die was done to provide a smooth extrusion and to reduce the load acting on die. The extrusion process was carried out in 100T hydraulic press. A tapered punch of 15 mm diameter and 50 mm length along with a 10 mm length taper similar to die's taper was provided at end of the punch. This punch was fixed at the top die of hydraulic press and then pressure was applied with lubrication oil to reduce friction between punch and die walls during extrusion process. The dimension of extruded specimen is 12 mm in diameter and 18 mm in length.

2.3 Material characterization

2.3.1 SEM micrographs

SEM micrographs of the powder mixture was to investigate the uniform distribution of nanoTiO_2 particulates in magnesium matrix. The SEM micrographs also to confirm agglomeration if any, particle size, grain and grain boundary refinement.

2.3.2 Density measurements

The mass density of extruded magnesium with 4, 8, 12, 16 and 20wt% TiO_2 was measured using Archimedes principle. Each sample was measured for 3 cycles to obtain accurate density values. Water was used as the standard liquid, which has a density of 1 g/cm^3.

2.3.3Microhardness test

Using a microhardness tester-Wolpert Group with Vickers indenter, the Microhardness tests were conducted on flat and polished extruded specimens. Each sample was measured for 3 cycles to measure the hardness accurately.

2.3.4 Compression test

The compressive properties of extruded Mg (4, 8, 12, 16 and 20wt%) TiO_2 samples were determined at room temperature. The test specimen of 12 mm diameter and 20 mm length was used. Each sample was tested 3 times to ensure repeatable values.

3. Results and discussions

3.1 SEM micrographs

SEM analysis was conducted on the elemental mixture of Mg-TiO_2 to confirm the uniform distribution of TiO_2 reinforcement over Mg matrix in five different weight percentage. Only if there is a uniform distribution, the bonding improves by grain growth and hence the properties of the resulting matrix would be considerably improved.SEM analysis was conducted on the extruded specimen to obtain the micrographic images of each composition of nanocomposites. As shown in Fig.3 the structure and grain growth of the nanocomposites continuously improved with increase in the addition of TiO_2 reinforcement. The grain size of pure magnesium is large and has a lot of micro pores [24]. But the addition of nano TiO_2 refined the grain structure. Mg-TiO_2 composite exhibits very smooth surface with few micro pores on its surface. It can be seen that the reinforcement is uniformly distributed in the matrix. The reason for uniform distribution of reinforcement particles is due to the efficient strategy adopted while fabrication of the nanocomposites. It was evident from the micrographs that TiO_2 distribution was quite homogeneous. Hot extrusion is the reason for grain growth in the magnesium matrix. The micrographs also exhibited the presence of minimal porosity in the developed nanocomposites with the result of density measurement through Archimedes principle.

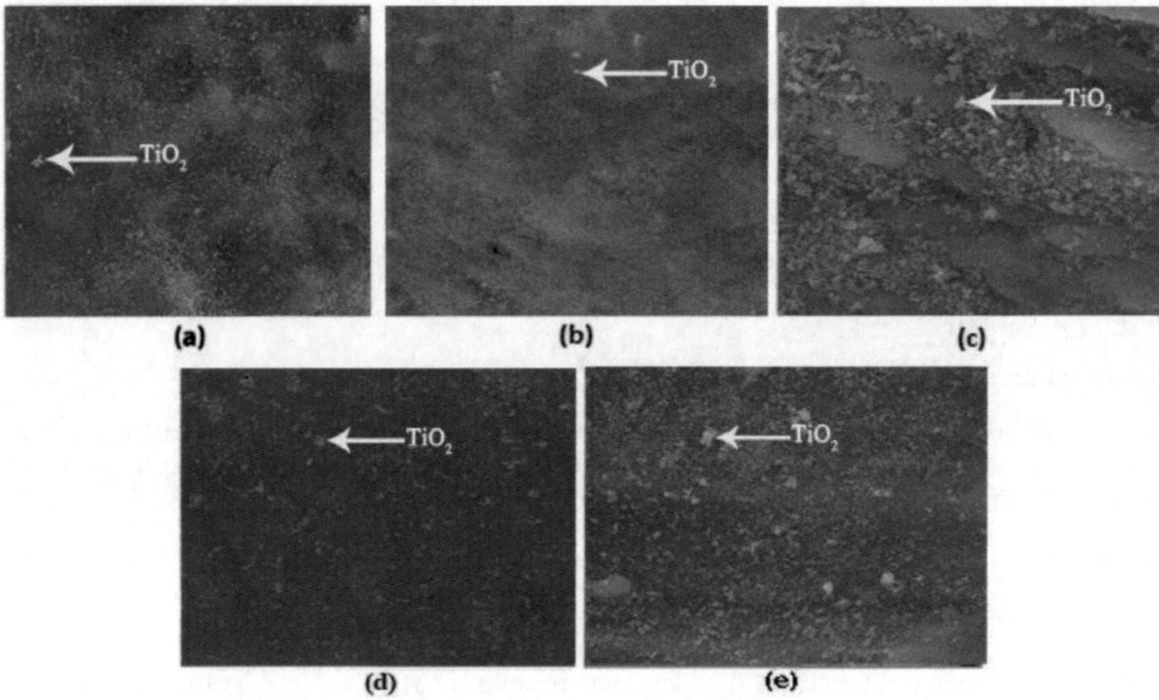

Fig.2. SEM micrographs of Mg-TiO_2 powder mixture (a)Mg-4%TiO_2, (b) Mg-8%TiO_2, (c) Mg-12%TiO_2, (d) Mg-16%TiO_2 and (e) Mg-20%TiO_2

3.2 Density and Porosity

Table 1. Results of density and porosity measurements.

S. No	Materials	Density		Porosity (%)
		Theoretical ρ (g/cm^3)	Experimental ρ (g/cm^3)	
1	Mg-4%TiO_2	1.839	1.837	0.1087
2	Mg-8%TiO_2	1.938	1.937	0.1012
3	Mg-12%TiO_2	2.039	2.037	0.1339
4	Mg-16%TiO_2	2.138	2.136	0.1183
5	Mg-20%TiO_2	2.239	2.236	0.1589

The presence of minimal porosity in the composite material can be due to better bonding between Mg-TiO_2 and the use of proper extrusion ratio. The results of density and porosity measurements conducted on the extruded specimen are shown in Table 1. Porosity is known to influence the mechanical properties to a great extent, therefore, pores must be eliminated through post sintering process and extrusion. The porosity of pure magnesium is higher when compared to reinforced magnesium [18]; the porosity level is low on an addition of 8 wt% of TiO_2 with Mg. The density of composites increased with the increase in weight percentage of TiO_2 particulates. The density features revealed minimal oxidation of magnesium and absence of macro pores.

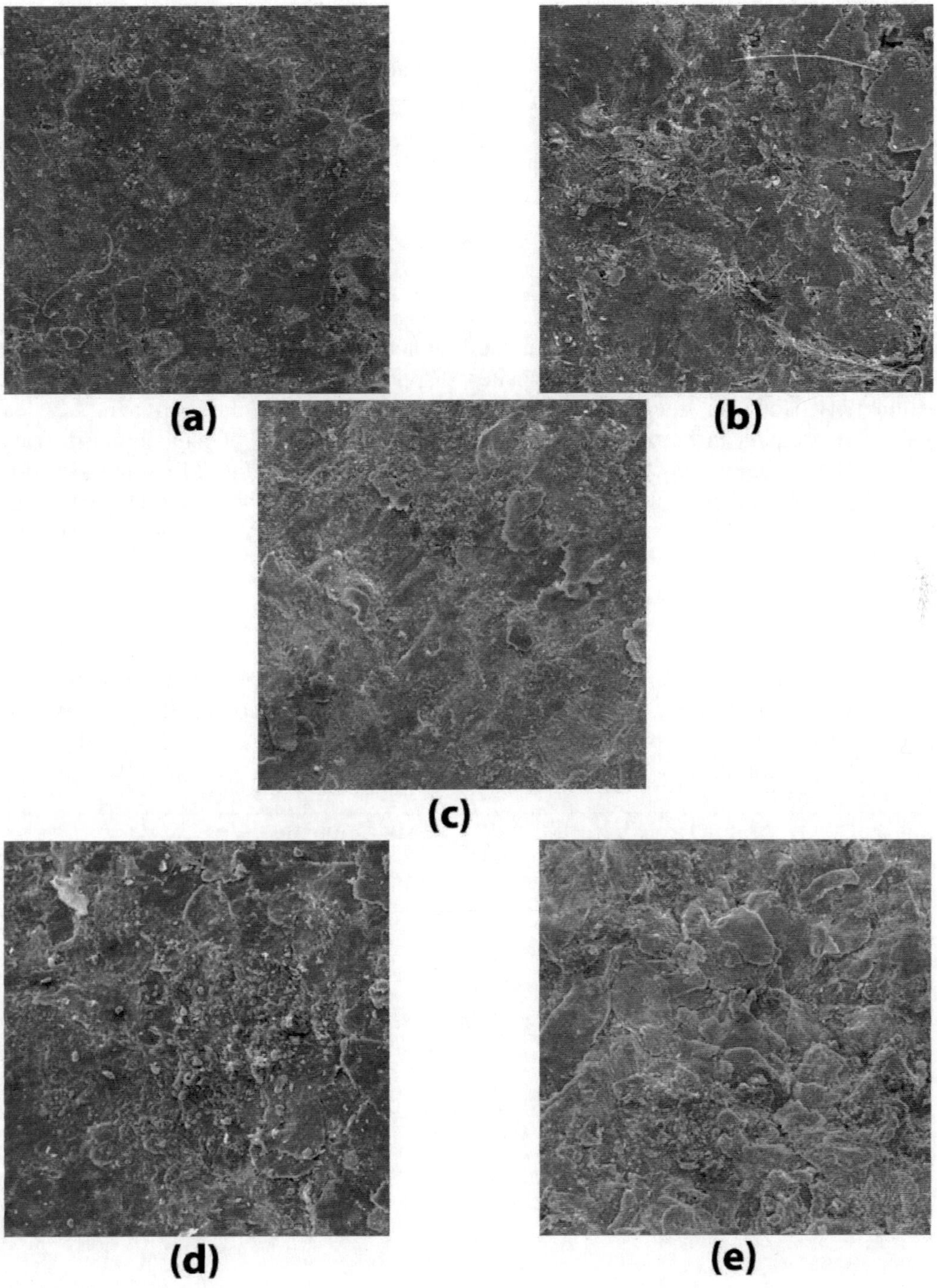

Fig.3. SEM Micrograph showing distribution of nano-TiO_2 particulates in Mg-TiO_2 composites: (a) Mg-4%TiO_2, (b) Mg-8%TiO_2, (c) Mg-12%TiO_2, (d) Mg-16%TiO_2 and (e) Mg-20%TiO_2.

3.3 Microhardness

Table 2. Result of hardness.

S. No	Composition	Average Hardness in HV
1	Mg-4%TiO_2	37
2	Mg-8%TiO_2	40
3	Mg-12%TiO_2	43.67
4	Mg-16%TiO_2	44.67
5	Mg-20%TiO_2	50

The room temperature mechanical properties are depicted in Table 2. It is found that pure Mg reveals very low hardness and the composites displayed higher hardness compared to pure magnesium [24]. However, the addition of nano-sized TiO_2 particulates to pure Mg leads to significant enhancement in hardness. The improvement in hardness of nanocomposites may be due to the strengthening effects that have arisen from the refined grains. The synergetic effect of TiO_2 particulates in magnesium matrix revealed improved hardness values. The value of microhardness continuously increased with increase in addition TiO_2 reinforcement. An intending load of 0.1 kg for a dwell time of 15 s was used.

3.4 Compression test

The compressive strength of extruded Mg (4-20 wt%) TiO_2 samples were tested at ambient temperature. The test specimens of diameter 12 mm of five different compositions were used. The specimen was subjected to compression test at a ram speed of 3 mm per min. The test was carried out on five different compositions of Mg (4-20 wt%) of TiO_2 nanocomposites.

Table3. Result of compressive strength.

S. No	Composition	Compressive strength in MPa
1	Mg-4%TiO_2	292
2	Mg-8%TiO_2	307
3	Mg-12%TiO_2	312
4	Mg-16%TiO_2	321
5	Mg-20%TiO_2	330

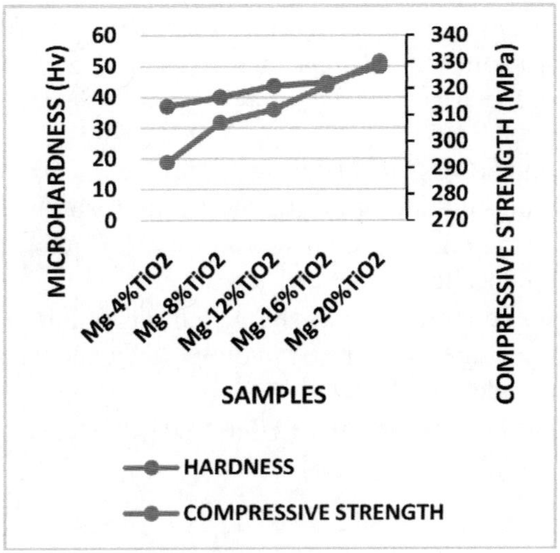

Fig.4. Graph correlating Microhardness and Compressive strength

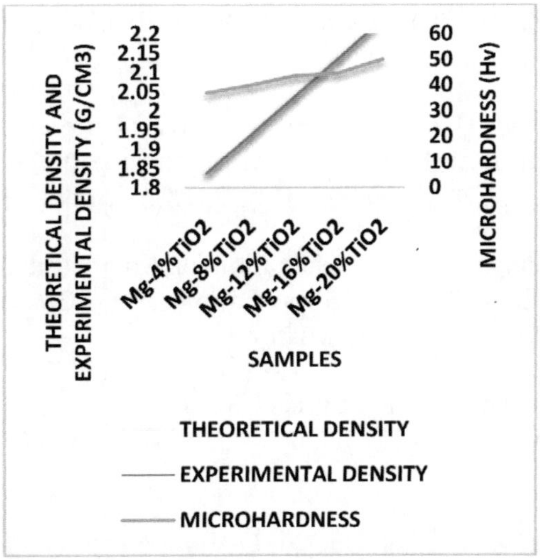

Fig.5. Graph correlating Theoretical density, Experimental density and Microhardness

The variation in microhardness and compressive strength of the composite samples complement each other. At higher weight fraction of reinforcement, in spite of original variation in it, there is a drastic increase in compressive strength. This may be due to the restricted grain growth at a higher content of reinforcement.

Density and hardness increase with respect to the quantity of reinforcement added to the matrix. This is attributed to the fact such as the non-uniform crystalline structure and abrasive nature of reinforcement.

4. Conclusions:

Mg-TiO_2 nanocomposites were successfully synthesized through powder metallurgy method followed by sintering and hot extrusion. The technology route follows as conversion of macro

IOP Conf. Series: Materials Science and Engineering **574** (2019) 012006 doi:10.1088/1757-899X/574/1/012006

titanium particles to nanoparticles, preparing the elemental mixture, compacting the elemental mixtures into cylindrical specimens, sintering the specimens, extruding the specimens, and characterizing mechanical properties. Based on microstructural and mechanical characterization following conclusions can be drawn.

- Powder metallurgy method is an efficient technique to fabricate Mg-based composite since the method prevents oxidation of Mg and less porosity is achieved.
- SEM image confirms the uniform distribution of TiO_2 over Mg matrix and thus proves homogeneity of the composite.
- Density is observed to be increasing as the weight percentage of reinforcement increases and porosity is minimum due to the fine bonding of Mg and TiO_2 in the mixture and hence proves the minimum oxidation of Mg.
- Hardness has been considerably improved due to the synergetic effect of TiO_2 particles as the weight percentage of TiO_2 increases.

References:

[1] Devaraju, A and Sivasamy,P. "Comparative Analysis of Mechanical Characteristics of Sisal Fibre Composite with and without Nano Particles", *Materials Today: Proceedings, Vol. 5 (*2018), pp.14362-14366.

[2] Devaraju, A, Babu, K and Gnanavel babu, A . "Investigation on the Mechanical properties of Coconut Bunch fiber Reinforced Epoxy with Al_2O_3 Nano particles Composites for Structural Application", *Materials Today: Proceedings, Vol. 5 (*2018), pp.14252-14257.

[3] Devaraju, A., and Pazhanivel, K. "Evaluation of Microstructure, Mechanical and Wear Properties of Aluminium Reinforced with Boron Carbide Nano Composite*", Indian Journal of Science and Technology*, Vol. 9(2), pp. 1-6, 2016.

[4] Jianfang Li, Xiangyu Zhang, Xiaojing He, Ruiqiang Hang, Xiaobo Huang, Bin Tang, Preparation, biocompatibility and wear resistance of microstructuredZr and ZrO2alloyed layers on 316L stainless steel, 2017.

[5] Sheida Shiri, Chunzi Zhang, AkindeleOdeshi, Qiaoqin Yang, Growth, and characterization of tantalum multilayer thin films on CoCrMo alloy for orthopedic implant applications, 2017.

[6] M.M. Machado López, J. Faure, M.I. Espitia Cabrera, M.E. Contreras García, Structural characterization and electrochemical behavior of 45S5 bioglass coating on Ti6Al4V alloy for dental applications, 2015.

[7] Wong Wai Leong Eugene, Manoj Gupta, Characteristics of Aluminium and Magnesium Based Nanocomposites Processed Using Hybrid Microwave Sintering, 2009.

[8] Junko Umeda, Masashi Kawakami, KatsuyoshiKondoh, EL-Sayed Ayman, Hisashi Imai, Microstructural and mechanical properties of titanium particulate reinforced magnesium composite materials, 2010

[9] Ganesh Kumar MeenakshiSundaram, MuiHoonNai, AbdulhakimAlmajid, Manoj Gupta, Development of high-performance Mg–TiO2nanocomposites targeting for biomedical/structural applications, 2014.

[10] Cheng-Jie Li, Hong-feiSun, Wen-bin Fang, Effect of extrusion temperatures on microstructures and mechanical properties of Mg-3Zn-0.2Ca-0.5Y alloy, 2014.

[11] ShahrouzZamaniKhalajabadi, Aminudin Bin Haji Abu, Norhayati Ahmad, Muhammad Azizi Mat Yajid, NorizahBtHjRedzuan, RozitaNasiri, WaseemHaider, ImanNoshadi, Bio-corrosion behavior and mechanical characteristics of magnesium titania-hydroxyapatite nanocomposites coated by magnesium-oxide flakes and silicon for use as resorbable bone fixation material, 2017.

[12] AlirezaVahid, Peter Hodgson, Yuncang Li, Reinforced magnesium composites by metallic particles for biomedical applications, 2017.

[13] Dongfang Zhang, Zhengbing Qi, Binbin Wei, Zhoucheng Wang, Low-temperature thermal oxidation towards hafnium-coated magnesium alloy for biomedical application, 2017.

[14] J.A. Calder´on, J.P. Jim´enez, A.A. Zuleta, Improvement of the erosion-corrosion resistance of magnesium by electroless Ni-P/Ni(OH)2-ceramic nanoparticle composite coatings, 2016.

[15] Y.L. Xi, D.L. Chai, W.X. Zhang, J.E. Zhou, Titanium alloy reinforced magnesium matrix composite with improved mechanical properties, 2005.

[16] S.F. Hassan, M. Gupta, Development of ductile magnesium composite materials using titanium as reinforcement, 2002.

[17] GuangyaoXiong, YanjiaoNie, Dehui Ji, Jing Li, Chunzhi Li, Wei Li, Yong Zhu, Honglin Luo, Yizao Wan, Characterization of biomedical hydroxyapatite /magnesium composites prepared by powder metallurgy assisted with microwave sintering, 2016.

[18] R. Anish, M. Sivapragash, G. Robertsingh, the Compressive behavior of SiC/ncsc reinforced Mg composite processed through powder metallurgy route, 2017.

[19] B. Selvam, P. Marimuthu, R. Narayanasamy, V. Anandakrishnan, K.S. Tun, M. Gupta, M. Kamaraj Dry sliding wear behavior of zinc oxide reinforced magnesium matrix nano-composites, 2014.

[20] Guangli Bi, Yuandong Li, Xiaofeng Huang, Tijun Chen, Ying Ma, Yuan Hao, Dry sliding wear behavior of an extruded Mg-Dy-Zn alloy with long period stacking ordered phase, 2014.

[21] Aniruddha Das, Sandip P. Harimkar, Effect of Graphene Nanoplate and Silicon Carbide Nanoparticle Reinforcement on Mechanical and Tribological Properties of Spark Plasma Sintered Magnesium Matrix Composites, 2014.

[22] Kai Wen, Ke Liu, Zhaohui Wang, Shubo Li, Wenbo Du, Effect of microstructure evolution on mechanical property of extruded Mg-12Gd-2Er-1Zn-0.6Zr alloys,2014.

[23] Said Murat Kayhan, Aydin Tahmasebifar, MuammerKoc¸, Yusuf Usta, Ays¸enTezcaner, ZaferEvis, Experimental and Numerical Investigations for Mechanical and Microstructural Characterization of Micro-Manufactured AZ91D Magnesium Alloy Discs for Biomedical Applications, 2015.

[24] Muhammad Rashad, Fusheng Pan, Muhammad Asif, Jia She, Ahsan Ullah Improved

IOP Conf. Series: Materials Science and Engineering **574** (2019) 012006 doi:10.1088/1757-899X/574/1/012006

mechanical proprieties of "magnesium-based composites" with titanium-aluminum hybrids, 2014.

[25] Ayman Elsayed, KatsuyoshiKondoh, Hisashi Imai, Junko Umeda, Microstructure and mechanical properties of hot extruded Mg-Al–Mn–Ca alloy produced by rapid solidification powder metallurgy, 2009.

[26] R. del Campo, B. Savoini, A. Muñoz, M. A. Monge, R. Pareja, Processing and mechanical characteristics of magnesium-hydroxyapatite metal matrix biocomposites, 2016.

[27] Ayman Elsayed, Junko Umeda, KatsuyoshiKondoh, the texture and anisotropy of hot extruded magnesium alloys fabricated via rapid solidification powder metallurgy, 2011.

Experimental and Numerical Study on Modified Vertical Axis Wind Turbine

A.Abraham Eben Andrews [a]*, Karthick.P[a], A.Allan Jeraled[b], Anuroop.K[a], Chidanand. G[a]

[a]*Department of Mechanical Engineering, Hindustan Institute of Technology and Science, Hindustan University, Chennai, TamilNadu, India.*

[b]*Marg Institute of Design and Architecture Swarnabhoomi, MARG Swarnabhoomi, Velur Village Cheyyur post, Kanchipuram Dt, Tamilnadu, India*

Corresponding author: abrahamebenandrews@gmail.com

Abstract: Wind energy is very important in the development of sustainable energy with the growing focus on renewable energy; interest in design of wind turbines has also been expanding. In today's market, the horizontal axis turbine is the most common type in use; Experimental data of this research is to compare the CFD Analysis with the Experimental Data of hybrid Modified vertical axis wind turbine. The graphical data based on the parameters such as Velocity, Power Coefficient, Tip Speed Ratio Vs Power Coefficient and Wind Velocity Vs Power Coefficient, which are discussed in this research. The Savonius turbine (Lift Turbine) was attach with the Gorlov turbine (Drag driven turbine) for the self-starting of the hybrid-modified turbine.Here the design, construction and testing of vertical axis wind turbine with three blades, starting solely from low energy of the wind. The velocities such as 5, 7, 9 and 11 m/s are tested in CFD Analysis for checking from which maximum power can be extracted so it can be used in desert area.

1. Introduction

In present days renewable energy is motivated to use due to the increase in environmental problems mainly focusing on the global warming, windenergy plays a vital role in the renewable energy In India renewable fuel requires many technological improvements [1]. Horizontal Axis Wind Turbines (HAWT) and Vertical Axis Wind Turbines (VAWT) are the classifications of wind turbine. In many countries for medium-to-large scale Energy generations Horizontal axis wind Turbine is used and it isnot recognized as a viable option to harness the energy of the wind in urban areas, where the wind is less intense, much more chaotic and turbulent. Vertical axis Wind Turbine are suggested for large scale deployment as a better choice for cities areas because of their good performance even in weak and unstable wind[2]. To generate the Lift aerodynamic bodies like airfoil required in darries type wind turbine. So darries is costlier than Savonius type wind turbines. Savonius Vertical axis wind turbine was named after its inventor Siguard. J. Savonius [3]. The performance analysis of Savonius turbine is done by analysing the importance of its geometrical parameters like overlap ratio, aspect ratio, gap ratio and shape parameters [4].

Vertical axis wind turbineas shown in Figs. 1 combines the working principle of Drag and Lift type vertical axis wind turbine the solid model of this turbine is created in ANSYS Fluent workbench [11]and CFD analysis is carried out the suitable design is fabricated and tested in wind tunnel.The raw data measured during the open environment test on this wind turbine are wind speed (V),rotational speed of the rotor shaft and load applied on the dynamometer.

These data are processedand presented as per the wind turbine research community approach also provides the aerodynamics performance of the Wind Turbine in terms of dimensionless parameter as coefficient of power (Cp) and coefficient of torque (CT). Further to disclose the influence of tip speed ratio (TSR) on aerodynamic performance, the variations ofTSR with Cp and CT at different wind velocities [9]. Rotor showed its peak performance at an overlap of 0.16. Sukanta Roy andUjjwalSaha [4] tested modified Bach typevertical axis wind turbine design and the results were compared with normal types.Kamojiet al [5] studied thetorque formation in negative way in simple Savonius rotors and found that simple 90 degree twist can overcame this with a helical Savoniusrotors. Damak [6] tested helical Savonius rotors with 180º twist in a wind tunnel todetermine aspect ratio. Sukanta Roy and UjjwalSaha [7] tested the modifiedSavonius in a wind tunnel and the new design shows good performance. KhandakarNiazMorshed et al. [8] analysed three bladed Savonius rotor by numerically and experimentally to predict overlap ratio is optimum. The authors found that overlap ratio of 0.2 suits the hydroturbine. Muscoloa and ReziaMolfinob [10] developed a new blade design named Broznius rotors.

2. Designof Wind Turbine

2.1 Selection of Wind Turbine

As there are many types of vertical axis turbines with various efficiencies, life time and cost. Gorlov turbine has been selected from the following table based on the following parameters [12].

Table 1 Performance Comparison of Vertical Turbines

	H ROTOR	TURBY	GORLOV
Type (VAWT)	Darrieus	Darrieus	Darrieus
Swept Area (m2)	7.29	N/A	N/A
Cut-in wind speed (m/s)	3.9	4	5
Cut-out wind speed (m/s)	N/A	14	26
Rotor Diameter (m)	2.7	1.9	3.1
Weight (kg)	135	136	N/A
Expected Lifetime (years)	20	20	25
Annual energy production (kWh/yr)	2000-4000	N/A	4197 @ 5 m/s, 7000 @ 7 m/s
Rated power (kW)	1.7	2.5	6.5
Total height (m)	7.9/12.5/18.6	6	13

2.2 Selection of Blade Angle and No of Blades

Number of blade selected is 3 and at an angle taken is 58.37 degree [13]. The table show the values of slodity ration of both the blades in their same angle (58.37 degree) approximately same.

IOP Conf. Series: Materials Science and Engineering **574** (2019) 012007 doi:10.1088/1757-899X/574/1/012007

Table 2 Various Parameters of Turbine

Φ(degree)	N	H(m)	D(m)	C(m)	Σ
43.68	3	0.75	0.75	0.15	0.191
51.85	3	0.8	0.6	0.15	0.239
58.37	3	0.85	0.5	0.15	0.286
47.5	3	0.8	0.7	0.15	0.205
51.85	4	0.75	0.75	0.12	0.204
59.5	4	0.8	0.6	0.12	0.255
65.2	4	0.85	0.5	0.12	0.306
55.5	4	0.8	0.7	0.12	0.218

2.3 Wind Turbine Design Parameters

The wind turbine parameters considered in the design process are:

For Gorlov Turbine:

Height of Turbine, H = 0.8m, Diameter of Turbine, D = 0.5m
No. of Blades, N = 3
Chord Length, C = 0.15m
Density of Air, ρ = 1.225 kg/m3
Swept Area, A= D x H =0.8×0.5 = 0.4m2
Solidity, σ = (NC) ÷ (2πr) = (3×015) ÷ (2×3.14×0.25) = 0.2864
Aspect Ratio = H÷D = 0.5÷0.8 = 1.666

For Savonius Turbine:

Height of Turbine, H = 0.25m, Diameter of Turbine, D = 0.5m
Diameter of Blade, d =0.3m
Eccentricity, e = 0.1m
Aspect Ratio = H ÷ D = 0.5

3. Construction of Wind Turbine

A shaft of length 1800mm is taken and it has been converted into a stepped shaft from one end of length 250mm and a diameter of 20mm is kept and the next 300mm length is made into 25mm diameter and from the other end a length of250mm is made into a diameter of 25mm the rest of the shaft is kept at 30 mm diameter. Hybrid turbine made out of fibre-reinforced material so it is light in weight and can bemade to rotate [14]. Height of the Savonius turbine is 250mm since it is used only for self-starting and the height of the Gorlov is 800 mm, diameter of both the turbine is 500mm and the no of blades used is 3 in Gorlov turbine because of the construction becomes simpler and the cost factor and also the solidity ratio can be managed. The helical blade in Gorlov kept at 120-degree angle between the starting and ending point of the blade. Is mounted on stand and placed in front of Axial Flow Turbine (wind tunnel)[17].

IOP Conf. Series: Materials Science and Engineering **574** (2019) 012007 doi:10.1088/1757-899X/574/1/012007

3.1 Specifications of Axial Flow Turbine

Height = 1m
Minimum Speed = 400rpm
Maximum Speed = 982rpm
Discharge = 11m/s

Fig. 1 Solid Work Model of Hybrid Vertical Axis Wind Turbine

Fig 2 Turbine mounted in the axial flow wind tunnel

4. CFD Analysis of Hybrid Vertical Axis Turbine

4.1 Simulation Properties & Conditions

Table 3 Conditions for CFD Analysis

Time scheme	Transient state
Material	Air
Properties of Fluid	
Density:	1.225 kg/m3
Kinematic viscosity:	1.7894e-5 kg/m-s
Compressibility:	Incompressible
Cell zone conditions	
Continuum type 1:	Fluid (blade rotational region)
Continuum type 2:	Solid (Turbine tower)
Blade rotational velocity	200 rpm
Blade rotational direction	Anti-Clockwise
Boundary Conditions	
Inlet Velocity	5m/s, 7m/s, 9m/s, 11m/s
Atmospheric pressure	1.01325 Bar
Turbine	Rotational, relative to the adjacent cell zone
Wall	Stationary wall, specified shear is zero

IOP Conf. Series: Materials Science and Engineering **574** (2019) 012007 doi:10.1088/1757-899X/574/1/012007

4.2 Domain Details and Approach

The turbine is modeled in a generic domain to match the wind tunnel testing conditions in Fig 3. The length and width of the domain is 15m and 6m, and the height of the domain is approximately 12 times of the rotor [18].

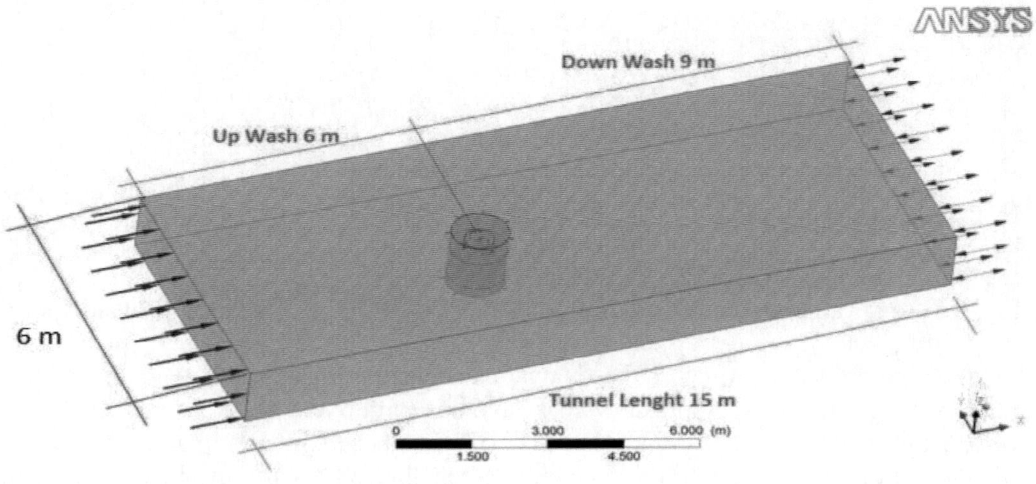

Fig. 3 CFD domains Configuration

4.3 Torque Variation for Velocities

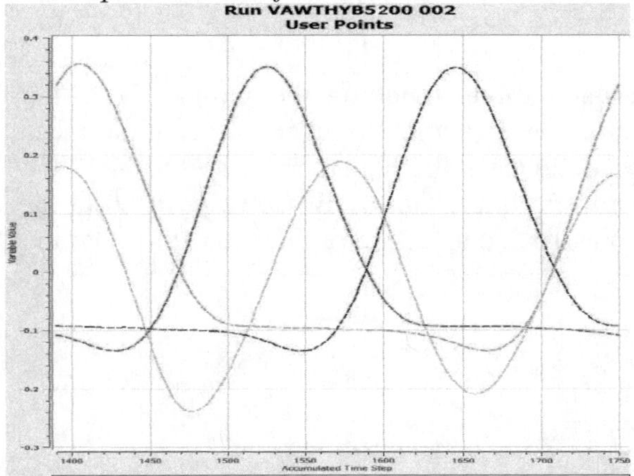

Fig. 4 Torque variation for velocities at 5m/s

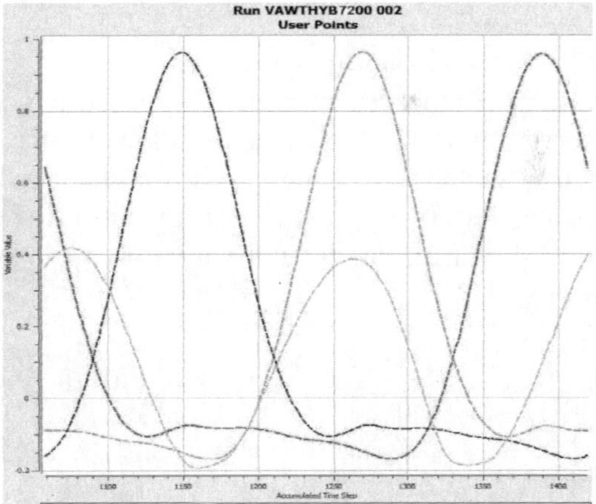

Fig. 5 Torque variation for velocities at 7m/s

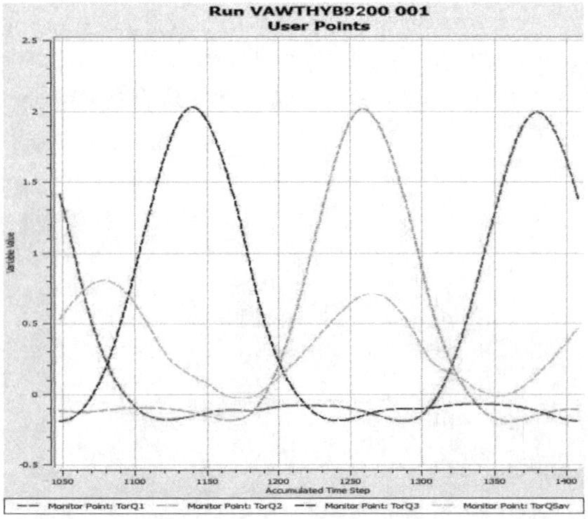

Fig. 6 Torque variation for velocities at 9m/s

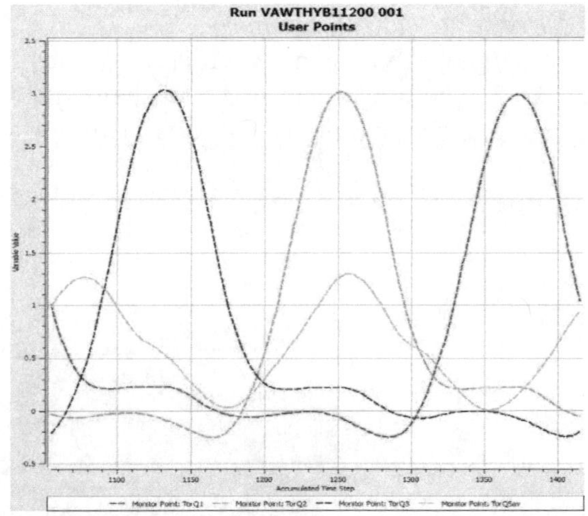

Fig. 7 Torque variation for velocities at 11m/s

This graph indicates the torque produced at varies angle of contact of the wind on the blade in fig 4,5,6,7. Yellow line on the graph shows the Savonius torque, the graph goes from top to bottom due to the shape Savonius turbine [16]. The Gorlov has 3 blades in it each blade gives a particular torque red line represents blade 1, green line represents the blade 2 and blue line represent blade 3. When the wind hits the blade 1, its observed that red line reaches the peak value. Similarly, when the wind hits the blade 2 and 3 it reaches its own peak value in the graph. These values of the peak are taken after leaving for few angles, so that it attains the constant values of the torque.

4.4 Pressure Variation of Turbine

Pressure contour represents the pressure values at various point on the turbine. Here the observation shows higher zone to lower zone of pressure (we know that the flow of fluid takes place when there is a pressure difference) red zone indicates the higher pressure and blue line indicates the lower pressure.forSavonius is shown in fig 8 and Gorlov is shown in fig 9 So the wind flows from higher pressure zone to lower pressure zone so the turbine start to rotate the turbine.

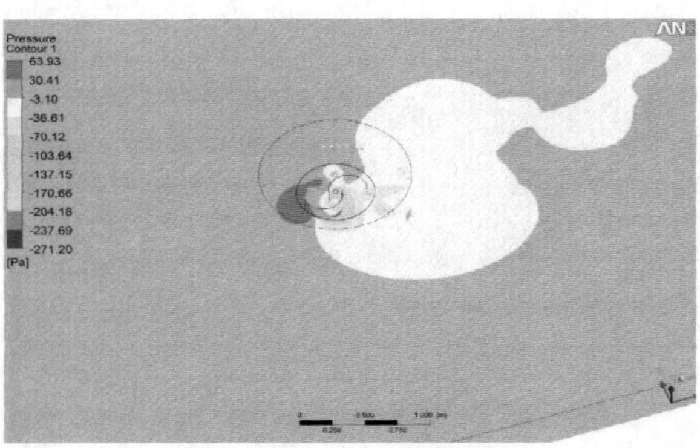

Fig. 8 Savonius Turbine

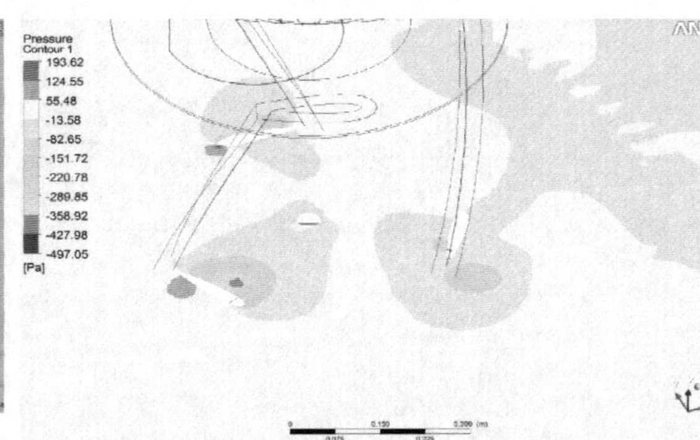

Fig. 9 Gorlov Turbine

4.5 Velocity Variation of Turbine

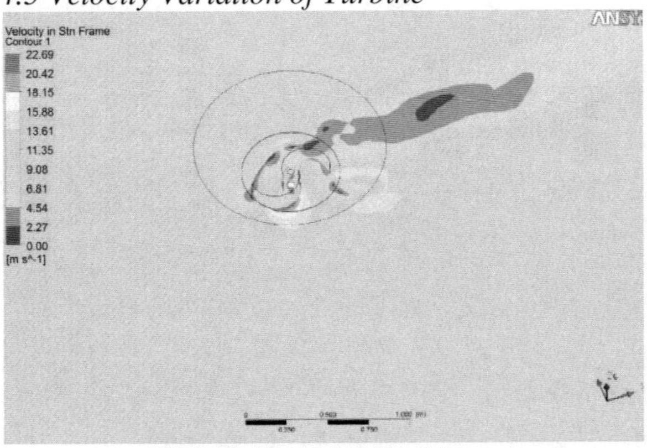

Fig. 10 Savonius Turbine

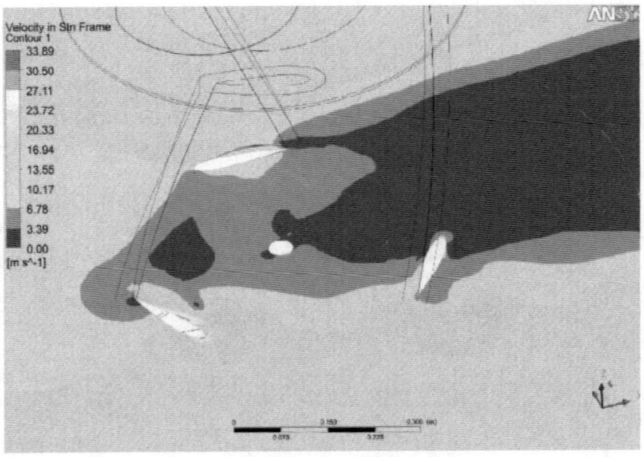

Fig. 11 Gorlov Turbine

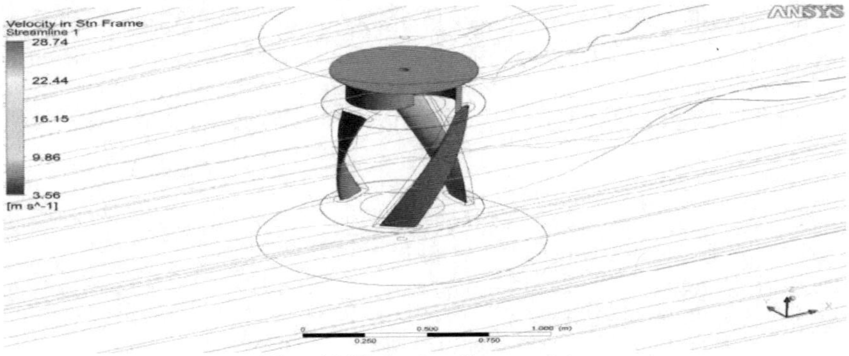

Fig. 12 Velocity Stream Line

Velocity stream line shows in fig 12 how the wind transmits it energy into the turbine for it to rotate[15]. Stream line shown in diagram, close observation gives the information that before hitting on the blades it has lost the kinetic energy while hitting on the blade and it is converted into the pressure energy. This will make the turbine to rotate there by the exit wind will have some loss in kinetic energy.

FOR 9 m/s

Torque = 1.358488 N-m (From Graph)

Speed N = 200 rpm

Angular velocity, $\omega = 2\pi N \div 60 = 20.9439$ rad/s

Tip Speed Ratio, $TSR = \omega r \div v = 1.1635$

$P_{in} = 0.5\rho AV^3 = 234.419$ watt

$P_{out} = T \times \omega = 28.4520$ watt

$C_p = P_{out} \div P_{in} = 0.1213$

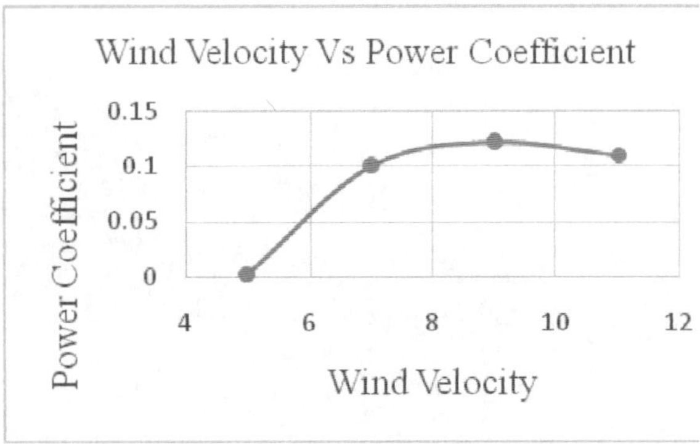

Fig.13 Wind Velocity Vs Power Coefficient

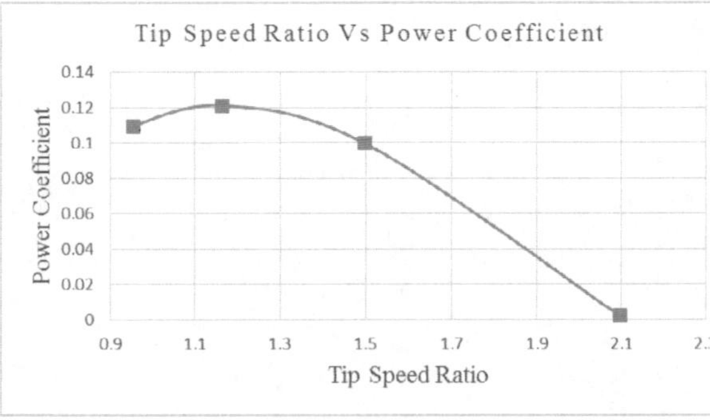

Fig. 14 Tip Speed Ratio Vs Power Coefficient

Fig 13 represents when the wind velocity increases the coefficient of power increases gradually and it reaches maximum and wind velocity further increases the coefficient of power decreases .This shows that at particular velocity only we can attain maximum coefficient of power and Fig 14 represents when the tip speed ratio increases the coefficient of power increases gradually and it reaches maximum and tip speed ratio further increases the coefficient of power decreases .This shows that at particular velocity only we can attain maximum coefficient of power.

5. Experimental Analysis

Calculation for 9 m/s

Torque = 0.07418 N-m (From Readings); Speed N = 201

Angular velocity, $\omega = 2\pi N \div 60 = 21.04$ rad/s

Tip Speed Ratio, TSR = $\omega r \div v = 1.1693$

Pin = 0.5ρAV3 = 234.419 watt

Pout = $T \times \omega = 1.56074$ watt, Cp = Pout÷Pin = 0.0066

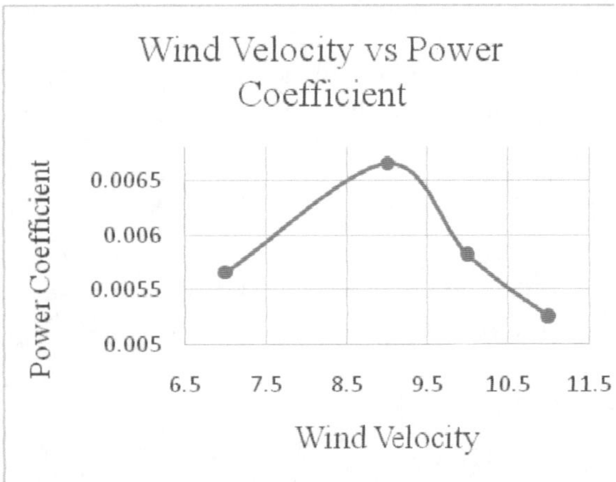

Fig. 15 Wind Velocity Vs Power Coefficient

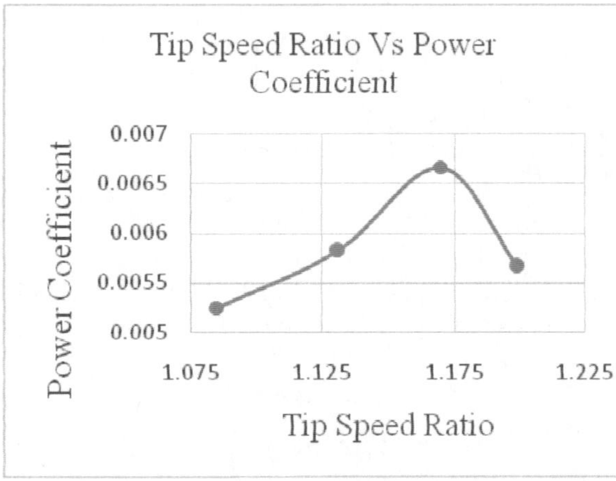

Fig. 16 Tip Speed Ratio Vs Power Coefficient

The above graph represents when the wind velocity increases the coefficient of power increases gradually and it reaches maximum and wind velocity further increases the coefficient of power decreases .This shows that at particular velocity only we can attain maximum coefficient of power. The above graph represents when the tip speed ratio increases the coefficient of power decreases gradually. This shows that at particular velocity only we can attain maximum coefficient of power.

5.1 Comparison of CFD and Experimental Analysis

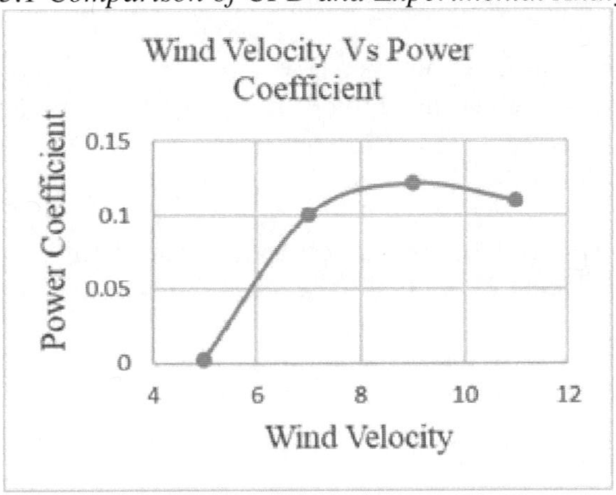

Fig. 17 Wind Velocity Vs Power Coefficient obtained from CFD analysis

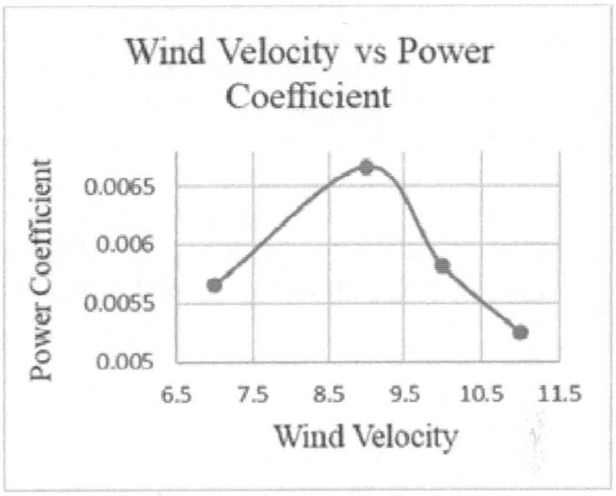

Fig. 18 Wind Velocity Vs Power Coefficient obtained by Experimental Analysis

Table 4 Comparison of Power Coefficient

Wind Velocity	CFD's Power Coefficient	Experimental Power Coefficient
7m/s	0.1001	0.00566
9m/s	0.1213	0.00666
11m/s	0.121	0.00525

Reasons for the Difference in Power Coefficient

1. In Experimental Analysis many losses to be considered such as
   - Frictional loss
   - Mechanical loss
   - Environmental loss
2. In CFD Analysis the turbine runs with some boundary conditions where friction and other mechanical losses are neglected.
3. The difference in power coefficient depends on manufacturing of blade (aerofoil).
4. Blower efficiency and losses.

6. Conclusion and recommendation

The various results observed in experimental and Numerical study, which is based on various parameters such as speed, torque, velocity and coefficient of power. The values obtained are as follows:

IOP Conf. Series: Materials Science and Engineering **574** (2019) 012007 doi:10.1088/1757-899X/574/1/012007

Table 5 Observed Results

Experimental		Numerical	
Speed	215 rpm	**Speed**	200 rpm
Torque	0.07418 N-m	**Torque**	1.358488 N-m
Velocity	9 m/s	**Velocity**	9 m/s
Power Coefficient	0.00666	**Power Coefficient**	0.1213

The various coefficient of power obtained from the graphs, the analysis reveals that the velocity at 9m/s gives more efficient power for the particular design so this kind of turbines can be used in desert area.

References

[1] Eduardo JoséNovaes Menezes, Alex MaurícioAraújo, NadègeSophieBouchonneau da Silva, A review on wind turbine control and its associated methods, Journal of Cleaner Production 174 (2018) 945-953

[2] Rakesh Kumar, KaamranRaahemifarb, Alan S. Fung, A critical review of vertical axis wind turbines for urban applications, Renewable and Sustainable Energy Reviews 89 (2018) 281–291

[3] J. Kumbernuss, J.Chen, H.X. Yang, L. Lu. Investigation into the relationship of the overlap ratioand shift angle of double stage three bladed vertical axis wind turbine (VAWT), 107-108 (2012) 57-75.

[4] Sukanta Roy, Ujjwal K. Saha. Wind tunnel experiments of a newly developed two-bladedSavonius-style wind turbine, Applied Energy 137 (2015) 117–125.

[5] M.A. Kamoji, S.B. Kedare, S.V. Prabhu. Performance tests on helical Savonius rotors, Renewable Energy 34 (2009) 521–529.

[6] A. Damak, Z. Driss, M.S. Abid. Experimental investigation of helical Savonius rotor with a twist of 180, Renewable Energy 52 (2013) 136-142.

[7] Sukanta Roy, Ujjwal K. Saha. Wind tunnel experiments of a newly developed two-bladed Savonius-style wind turbine, Applied Energy 137 (2015) 117–125.

[8] KhandakarNiazMorshed, Mosfequr Rahman, Gustavo Molina and Mahbub Ahmed, Wind tunnel testing and numerical simulation on aerodynamic performance of a three-bladed Savonius wind turbine, International Journal of Energy and Environmental Engineering 4 (2013)18. 556 Mechanical Engineering Design

[9] Sarathkumar R, MichaPremkumar T, Seralathan S, Mohan T, Numerical Investigation of Modified Bach Type Vertical Axis Wind Turbine, *Applied Mechanics and Materials ISSN: 1662-7482, Vol. 852, pp 551-557*

[10] Giovanni Gerardo Muscoloa, ReziaMolfinob, From Savonius to Bronzinus: a comparison among vertical wind turbines, Energy Procedia 50 (2014) 10-18.

[11] ANSYS Training guide – ANSYS customer portal

[12] Josh Anderson, Bronwyn Hughes, Celestine Johnson, Nick Stelzenmuller, Leonardo Sutanto, Brett Taylor, design and manufacture of a cross flow helical tidal Turbine, june 14 2011

[13] Craig Stouta, Sheikh Islama and et al , Efficiency Improvement of Vertical Axis Wind Turbines with an Upstream Deflector, Energy procedia 118 (2017)141- 148

[14] MeltemAltinKaratas, Hasan G okkaya, A review on machinability of carbon fiber reinforced polymer (CFRP) and glass fiber reinforced polymer (GFRP) composite materials , Defence Technology (2018) 1- 9

[15] Tong Zhou, DietmarRempfer. Numerical study of detailed flow field and performance of Savonius wind turbines, Renewable Energy 51 (2013) 373-381.

[16] M.H. Nasef, W.A. El-Askary, A.A. AbdEL-hamid, H.E. Gad, Evaluation of Savonius rotor performance: Static and dynamic studies. J. Wind Eng. Ind. Aerodyn.123 (2013)1-11.

[17] Wei-ChengWang, Wen TongChong, Tien-HsinChao, Performance analysis of a cross-axis wind turbine from wind tunnel experiments, 174 (2018) 312–329

[18] John Kaiser Calautitn , Ben Richard Hughes Data in Brief 7 (2016)216-220.

IOP Conf. Series: Materials Science and Engineering **574** (2019) 012008 doi:10.1088/1757-899X/574/1/012008

Optimization through NSGA-II during machining of A356Al/20%SiCp metal matrix composites using PCD Tool

M. Seeman[a*], D. Kanagarajan[b], P. Sivaraj[c], R. Seetharaman[d] and A. Devaraju[e].

[a, b, c] Department of Manufacturing Engineering, Annamalai University, Annamalainagar - 608 002, Tamilnadu, India.
[d] Department of Mechanical Engineering, University College of Engineering, Anna University, Dindigul - 624 622, Tamilnadu, India.
[e]Professor, Department of Mechanical Engineering, Adhi College of Engineering and Technology,Kanchipuram-631605,Tamilnadu,India.

[*]Corresponding author Email: mfgseeman@gmailmail.com,

seeman_au575@rediffmail.com

Abstract

Metal matrix composite (MMC) has established growing usages in the engineering for lightweight high-strength application. However, because of abrasive nature of the reinforcement particles in MMC, machinability is reduced, tool wear is high, yet only diamond tools are appropriate for machining MMC. Being a complex process, it is very complicated to establish optimal parameters for improving cutting performance. Tool flank wear (*VBmax*) and Average surface roughness (*Ra*) are the most significant output responses, which decide the machinability of a material. The effect of cutting parameters on output responses of tool flank wear and average surface roughness are conflict with each one another so there is no single optimal mixture of cutting parameters. In this study multi regression model function, based on NSGA-II was used to optimize the machining of Al/20%SiCp MMC using tipped polycrystalline diamond (PCD) tool and also characterize the relationship among input parameters and output performance. The NSGA-II based non dominated solutions of 30 combinations chosen from 100 and presented; none of them superior to any other each and every one are best based on engineer requirement.

Keywords: metal matrix composite (MMC); tool flank wear (*VBmax*); average surface roughness (*Ra*); polycrystalline diamond (PCD); non-dominated sorting genetic algorithm (NSGA-II)

1 Introduction

Many of modern technologies require material with unusual combinations of properties that cannot be met by the conventional metals, alloys, ceramics and polymeric materials. This is particularly true for materials that are required for aeronautical, marine and automobile industries. The present day requirements are such that no material can satisfy all of them hence the new materials have to be synthesized. This resulted the Metal Matrix Composite (MMC) are receiving more attention. Among the MMCs Particulate Metal-Matrix Composites (PMMCs) are of particular interest, because they exhibit superior ductility and lesser anisotropy than FRP and GFRP composites [1]. These materials might have a broad application, particularly for components, which are exposing to friction [2]. Several important applications of connecting rods, piston, piston rings, cylinder liners, bearings, turbine blades, impellers and space structures rings, etc. [3]. The commonly used matrix and reinforcements are magnesium, titanium, Aluminum and SiC, Al_2O_3. The various advanced manufacturing process available to produce near-net shaped MMC components, but component design and dimensional tolerance requirements, the need for machining cannot be completely eliminated [4].

The problem with MMCs is that they are difficult to machine material, due to the hardness and abrasive nature of the SiC or other reinforcing particles, e.g., 2400 Hv of SiC vs. 1800 Hv of tungsten carbide (WC) [5]. Diamond is exception, for case in point, which is approximately three to four times harder than hard metal and does not have a chemical tendency to react with the work material. That is the reason several researchers have suggested that polycrystalline diamond (PCD) tools are the only tool material that is capable of providing a useful tool life during the machining of Al/SiCp MMCs.

EL-GALLAB and SKLAD [6, 7] reported, in a study of wear mechanism of diamond tools in machining MMC, that abrasion/adhesion by reinforced particulates are the dominant wear mechanisms and severe machining conditions will result in chipping in PCD and peeling of CVD diamond coating. LIN et al. [8] observed the flank wear as the dominant mode of tool failure in machining Al-SiC MMC using PCD tool. ANDREWES et al. [9] investigated the tool wear in machining Al/SiC composites by diamond tools and found that tool wear involves two stages, one being initial flank wear caused by abrasion of hard particles, and the other combined adhesion-abrasion when the work materials start accumulatively adhering to the tool wear-land. PAULO DAVIM [10] reported that cutting velocity has the most influence on tool wear followed by cutting time and feed rate; with respect to surface finish feed rate observed to have greater influence.

The principles of bi-performance optimizations vary from single performance optimization. In bi-performance optimization, there are two objective functions, everyone have a diverse optimal solution. Mostly objectives differ with one another (optimizing one compromise another) [11]. The Genetic Algorithm is an evolutionary algorithm. It is based on the procedure of natural selection and it combines the individuality of direct search and probabilistic selection methods. GA is simple and potent tool for getting final optimal solutions for multi-model and combinational problems. Its work based on the principle of feasible solutions on population, therefore used in optimization of multi-objective problems to get solutions at the same time [12].

In the multi-objective optimization set of solutions superior then other solutions considering the all objectives. The inferior solutions compare to other solution in a search space only depend on one or more objectives. It is called non-dominated or Pareto-optimal solutions and other solutions are called as dominated solutions [11]. The method of multi-objective optimizations like GA has sufficiently established their use in finding a sound diverged and well-spread solution set of near Pareto-optimal solutions [13]. The NSGA-II has nature of high speed and elitist multi-objective GA mainly used producing Pareto-boundary. Important benefits of solve multi objective problems is that it leads the search toward the global Pareto front while maintaining diversity of the solution set along that front.

In this paper study the machining of 20%AlSiCp particulate metal matrix composites using tipped polycrystalline diamond (PCD) cutting tool. The turning experiments were conducted based on RSM design. Main cutting parameters like cutting speed (*V*), feed rate (*f*), depth of cut (*d*) and machining time (*t*) were optimized with respect to bi-performance character of flank wear (*VBmax*) and average surface roughness (*Ra*). Mathematical models were developed using RSM for tool flank wear and average surface roughness. The capability of the developed numerical models has been tested by the ANOVA test. With help of same models to optimize machining parameters using NSGA-II algorithm in the objective of minimizing tool flank wear and average surface roughness.

2 Experimental Study

Turning experiments were performed on a centre lathe which is having a spindle speed range of 30-1600 rpm with feed range 0.05-3.5 mm/rev with adequate spindle power. A356 (LM-25) aluminum alloy (7 Si, 0.33 Mg, 0.3 Mn, 0.5 Fe, 0.1 Cu, 0.1 Ni, 0.2 Ti) reinforced with green bonded silicon carbide average size of 25μm with a volume fraction of 0.20, which was manufactured through stir casting route was used as work material for carrying experimentation. The size of the work piece was 90 mm diameter and 250 mm length. The experimental setup for machining and A356 Al-20%SiCp MMC work piece are shown Figure 1&2. Considering the abrasive nature of work material tipped polycrystalline diamond (PCD) was used for turning which shown in Figure 3. The cutting speed (*V*), feed rate (*f*), depth of cut (*d*) and machining time (*t*) are consider as machining parameter, and the responses are tool flank wear (*VBmax*) and average surface roughness (*Ra*).

The experiments were designed using face centered Central composite design (CCD). The factorial portion of CCD is a full factorial design with all combinations of the factors at two levels (high, +1 and low, −1) and composed of the eight star points and seven central points (coded level 0) which is the midpoint between the high and low levels. The star points are at the face of the cubic portion on the design which corresponds to a α value of 1 and this type of design is commonly called the face centered CCD. Table 1 shows the levels of four machining parameters and their ranges. The experimental plans were carried out using the stipulated conditions based on the face centered CCD involving 31 runs in the coded form as shown in Table 2. The design was generated and analyzed using MINITAB statistical package. The turning experiments were carried out as per the conditions given by the design matrix at random to avoid systematic errors.

Table 1. Cutting parameter and their levels

Control Parameters	Unit	Symbol	Levels		
			-1	0	+1
Cutting speed	m/min	V	50	100	150
Feed rate	mm/rev	f	0.05	0.15	0.25
Depth of cut	mm	d	0.5	1.0	1.5
Machining time	min	t	2	4	6

The tool flank wear was calculated using CLEMAX optical microscope. To measure the tool wear the maximum width of flank wear land was considered. The average surface roughness (Ra) is generally practiced in industries, was measured in this study. The surface roughness was measured by using MITUTOYO SURF III surface tester.

Figure 1. Experimental setup for machining

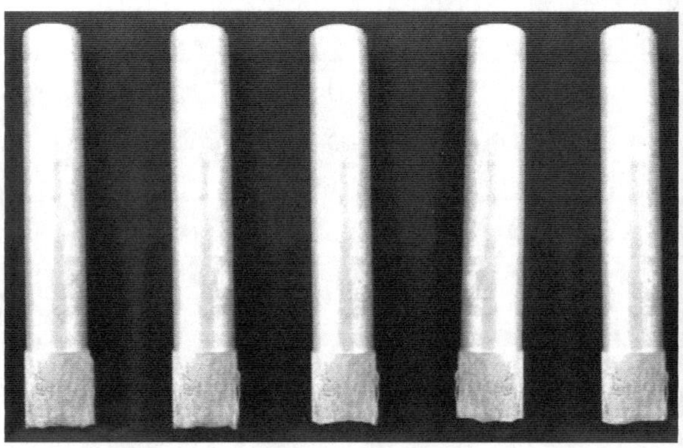

Figure 2. LM25 Al/20%SiCp MMC

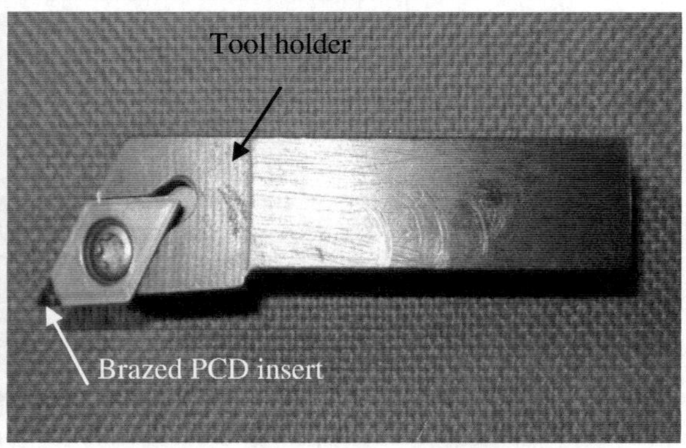

Figure 3. PCD tool with tool holder

3 Empirical modeling for flank wear and surface roughness

The higher order polynomial equations of empirical models for tool flank wear and surface roughness are developed using experimental results given in Table 2. The models given blow as follows:

$$\text{Tool Flank wear } (VB_{max}) = 0.06738 + 0.00077V - 0.31007f - 0.00184d + 0.00404t - 0.00000V^2 + 1.09747f^2 + 0.00190d^2 - 0.00013t^2 + 0.00027Vf + 0.00000Vd + 0.00001Vt + 0.00250fd + 0.00063ft + 0.00012dt \quad (1)$$

$$\text{Average surface roughness } (Ra) = 1.14619 - 0.00513V + 4.06338f - 0.00398d + 0.03814t + 0.00001V^2 - 2.61867f^2 + 0.0152\,d^2 - 0.00155t^2 - 0.00225Vf + 0.00005Vd + 0.00003Vt - 0.02500fd - 0.01250ft + 0.00250dt \quad (2)$$

Table 2. Experimental results

S.	Coded factors				Actual cutting parameters				Response variable	
No	X_1	X_2	X_3	X_4	*V*	*f*	*d*	*t*	Y_1 (VB_{max}, mm)	Y_2 (*Ra*, µm)
1	-1	0	0	0	50	0.15	1.0	4	0.099	1.58
2	1	0	0	0	150	0.15	1.0	4	0.170	1.21
3	0	-1	0	0	100	0.05	1.0	4	0.142	1.08
4	0	1	0	0	100	0.25	1.0	4	0.153	1.62
5	0	0	-1	0	100	0.15	0.5	4	0.135	1.36
6	0	0	1	0	100	0.15	1.5	4	0.139	1.40
7	0	0	0	-1	100	0.15	1.0	2	0.130	1.31
8	0	0	0	1	100	0.15	1.0	6	0.142	1.43
9	-1	1	1	1	50	0.25	1.5	6	0.126	1.94
10	1	-1	-1	-1	150	0.05	0.5	2	0.162	0.81
11	-1	-1	1	1	50	0.05	1.5	6	0.116	1.33
12	1	1	-1	-1	150	0.25	0.5	2	0.176	1.39
13	-1	-1	-1	1	50	0.05	0.5	6	0.113	1.28
14	1	1	1	-1	150	0.25	1.5	2	0.180	1.43
15	-1	-1	-1	-1	50	0.05	0.5	2	0.101	1.17
16	1	1	1	1	150	0.25	1.5	6	0.195	1.55
17	-1	1	-1	1	50	0.25	0.5	6	0.122	1.91
18	1	-1	1	-1	150	0.05	1.5	2	0.165	0.85
19	-1	1	-1	-1	50	0.25	0.5	2	0.106	1.81
20	1	-1	1	1	150	0.05	1.5	6	0.180	0.97
21	-1	1	1	-1	50	0.25	1.5	2	0.109	1.83
22	1	-1	-1	1	150	0.05	0.5	6	0.184	0.93
23	-1	-1	1	-1	50	0.05	1.5	2	0.103	1.21
24	1	1	-1	1	150	0.25	0.5	6	0.192	1.50
25	0	0	0	0	100	0.15	1.0	4	0.136	1.38
26	0	0	0	0	100	0.15	1.0	4	0.137	1.39
27	0	0	0	0	100	0.15	1.0	4	0.136	1.37
28	0	0	0	0	100	0.15	1.0	4	0.138	1.38
29	0	0	0	0	100	0.15	1.0	4	0.137	1.38
30	0	0	0	0	100	0.15	1.0	4	0.136	1.39
31	0	0	0	0	100	0.15	1.0	4	0.138	1.37

4 NSGA-II Optimization

The non-dominated genetic algorithm (NSGA-II) was used in this study to optimize the machining parameters in turning of 20%AlSiCp MMC using PCD tool. The main objectives of optimization are as given blow:

1) Minimization of tool flank wear (VB_{max})

2) Minimization of average surface roughness (*R*a).

The two-objective genetic algorithm optimization method used is a fast, elitist non-dominated sorting genetic algorithm (NSGA-II) developed by DEB et al [14]. This algorithm uses the elite-preserving operator, which favors the elites of a population by giving them an opportunity to be directly carried over to the next generation [15].

5 NSGA-II algorithm

The non-dominated sorting genetic algorithm has been criticized for its high computational complexity, lack of elitism and its choice of the optimal parameter value for sharing parameter σ. The NSGA-II is a modified version, which has a better sorting algorithm, incorporates elitism and does not require the choosing of a sharing parameter a priori. There are two key concepts in NSGA-II: a fast non-dominated sorting of the population and a crowding distance. The flow chart of the NSGA II program is shown in Figure 4.

5.1 Non-dominated sort

The initialized population is sorted based on non-domination. The non-domination is an individual and is said to dominate another if its objective function is no worse than the other and at least in one of its objective functions is better than the other. The fast-sort algorithm was described in Ref. [16].

5.2 Crowding distance

In NSGA-II, in addition to the fitness value, a new parameter called "crowding distance" is calculated for each individual. The crowding distance is a measure of how close an individual to its neighbors. Crowding distance is assigned front wise: comparing the crowding distance between two individuals in a different front is meaningless. The basic idea behind the crowding distance is finding the Euclidean distance between each individual in a front based on their m objectives in m-dimensional hyperspace. Initially, a random parent population of P_o, of size N is generated. The population is sorted based on non-domination level. Each solution is assigned a fitness level, and the best level is 1. Thus, minimization of fitness is assumed. Binary tournament selection, recombination and mutation operators are implemented to generate the child Q_o, of size N. The procedure for the remaining generation (for $t \geq 1$) can be found in Ref. [14].

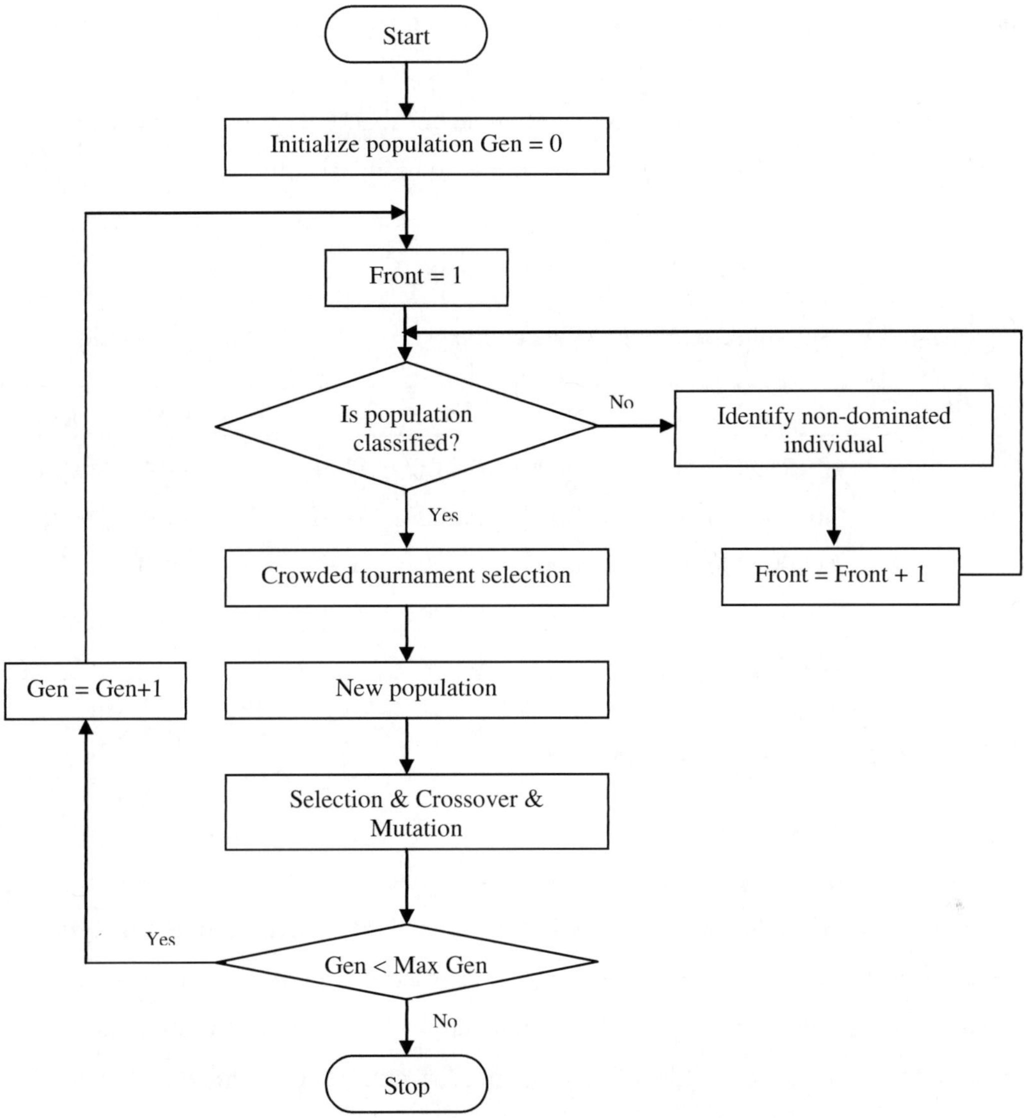

Figure 4. The flow chart of the NSGA II program

6 Results and discussion

The 20%AlSiCp MMC manufactured through stir casting method machining performance were studied using PCD cutting tool. Empirical model for flank wear and surface roughness developed by means of MINITAB software are given in Eq. (1) and (2). The ANOVA table for the quadratic model for VB_{max} & Ra is shown in Table 3. The standard percentage point of F distribution for 95% confidence limit is 4.06. From the Table 3 the F- values for VB_{max} and Ra are 3.30 & 3.15 respectively for lack-of-fit be smaller than the standard value. The lack of fit is insignificant as it is preferred. The other model terms are said to be significant. The value of R^2 calculated tool flank wear and surface roughness is 99.86% and 99.89% is higher than 99%, very close to unity, indicating a good correlation between the independent variables (factors) and the responses (VB_{max} & Ra). The P-value for the models is lower than 0.05 (i.e. $\alpha = 0.05$, or 95% confidence limit) which indicate that the model is measured to be statistically significant.

Table 3 Test results of ANOVA

Source	Degree of freedom	Sum of square		Adjusted mean square		F-Value		P-Value	
		VB_{max}	*Ra*	*VB_{max}*	*Ra*	*VB_{max}*	*Ra*	*VB_{max}*	*Ra*
Linear	4	0.02219	2.316	0.00009	0.0122	47.32	78.1	0.00	0.00
Square	4	0.00068	0.002	0.00017	0.0007	87.45	4.34	0.00	0.01
Interaction	6	0.00003	0.002	0.00001	0.0004	3.17	2.53	0.03	0.06
Lack of fit	10	0.00002	0.002	0.00000	0.0002	3.30	3.15	0.08	0.09
Error	6	0.00001	0.000	0.00000	0.0001				
Total	30	0.02295	2.324						
R^2 for VB_{max}= 99.86% and R^2 for Ra = 99.89%									

Apart from the above, the interaction effects are also analyzed and given in Table 4. From the table it is observed that the linear effect of cutting speed, feed rate and machining time, square effect of cutting speed, interaction effect between cutting speed and feed rate are more significant for tool flank wear, whereas other linear, square and interactions effects are not so significant. For surface roughness the linear, square effect of cutting speed, feed rate, liner effect of machining time, cutting speed interaction with feed rate are most significant, whereas other effects are not so significant as shown in Table 4. Figure 5 & 6 displays the normal probability plot of the residuals for tool flank wear and surface roughness. It can be noticed both VB_{max} and Ra residuals are located on a straight line, which means that the errors are normally distributed and the regression model is fairly adequate. Figure 5 (b) & 6 (b) indicates the maximum variation for tool flank wear and surface roughness are -0.0020 to 0.0015 and -0.03 to 0.03, which shows the high correlation that exists between fitted values and observed values.

IOP Conf. Series: Materials Science and Engineering **574** (2019) 012008 doi:10.1088/1757-899X/574/1/012008

Table 4 Estimated regression coefficients

Symbol	VB_{max}		Ra	
	Coefficient	**P-value**	**Coefficient**	**P-value**
Constant	0.06738	0.000	1.64619	0.000
V -m/min	0.00077	0.000<	-0.00513	0.000<
f -mm/rev	-0.31007	0.000<	4.06338	0.000<
d -mm	-0.00184	0.806	-0.00398	0.952
t -min	0.00404	0.043<	0.03814	0.033<
V^2	-0.00000	0.034<	0.00001	0.028<
f^2	1.09747	0.000<	-2.61867	0.004<
d^2	0.00190	0.593	0.01525	0.630
t^2	-0.00013	0.555	-0.00155	0.437
V f	0.00027	0.001<	-0.00225	0.002<
V d	0.00000	0.726	0.00005	0.694
V t	0.00001	0.094	0.00003	0.436
f d	0.00250	0.726	-0.02500	0.694
f t	0.00063	0.726	-0.01250	0.436
d t	0.00012	0.726	0.00250	0.436

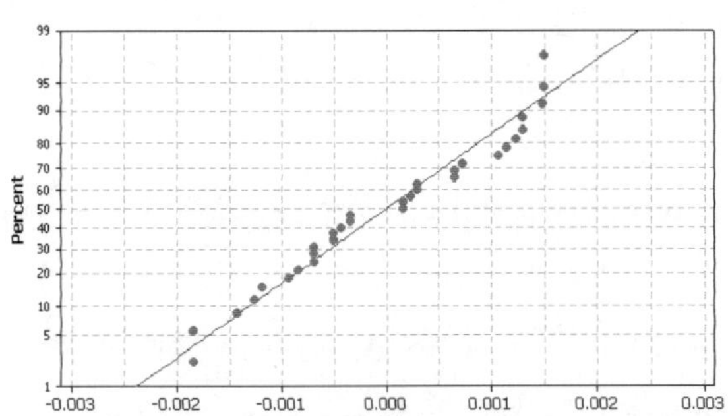

(a) Normal probability plot for VB *max*

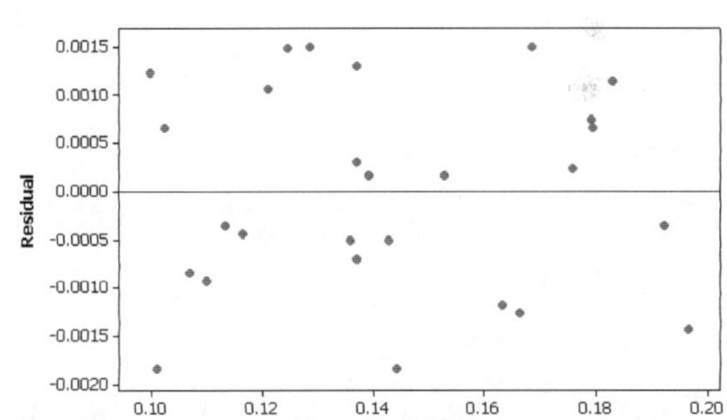

(b) Residual verses fitted value for VB *max*

Figure 5. Residuals plot used for tool flank wear (VB *max*)

A single objective optimization algorithm will generally be ended obtain an optimal solution. However, for the majority of the multi-objective problems, there could be a more number of optimal solutions. Appropriateness of one solution depends on a number of factors including user's preference and problem atmosphere, and hence discovery the complete set of optimal solutions may be preferred. Among the Pareto optimal result, none of the solutions is particularly superior to any other solution and hence this solution is extremely better than any other solution and therefore this solution is called as non-dominated solution.

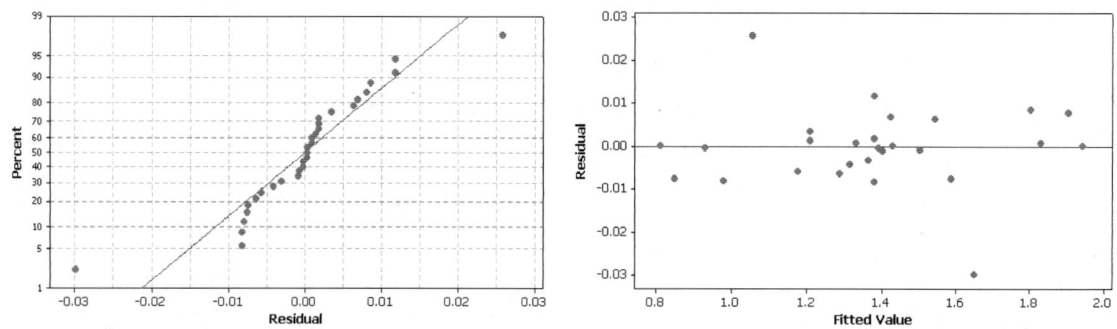

(a) Normal probability plot for Ra **(b) Residual verses fitted values for Ra**

Figure 6. Residuals plot used for average surface roughness (Ra)

Genetic algorithm is competent, adaptive, and healthy search and optimization technique, used to find solutions to linear and non-linear problems. Based on the principle of natural genetic system by simultaneously exploring multiple area of the solution space and exponentially exploit potential regions by means of selection, crossover and mutation. In genetic system information of each individual or possible solution is determined in structures called chromosomes. In common, the fittest individuals of population are more possible to reproduce and carry on to the next generation and improving succeeding generations. The Non-Dominating Sorting Genetic Algorithm (NSGA-II) developed by DEB in the year 2002 is finest method for generate the Pareto frontier and is used in this work.

The NSGA-II program positions the individuals as based on dominance. The fast non-dominated sorting process permits us to discover the non-domination frontiers where individuals of the frontier set are not dominated by any other solution. The crowding distance was calculated for every individual of the new population. Crowding factor gives the GA, the skill to differentiate individuals that have the same class. This forces the GA to homogeneously cover the frontier rather than bunch up at numerous high-quality points by trying to keep population variety. The comparison operator ($<_n$) is used by GA to sort the population for selection purposes.

The procedure was repeated ten times to obtain a larger number of points in the Pareto solution set. Non-dominated solution set obtained over the entire optimization procedure is revealed in Figure 7. This indicates the arrangement of the Pareto front leading to the finishing set of solutions. The matching goal function values and decision variables of this non-dominated solution set are given in Table 5. The 30 out of 100 sets be presented since no one of the solutions in the non-dominated set is absolutely better than any other; any one of them is an satisfactory solution. The selection of individual solution over another depends on the option of the process engineer. Either less tool wear or a good surface roughness is required; an appropriate mixture of parameters can be chosen from Table 5.

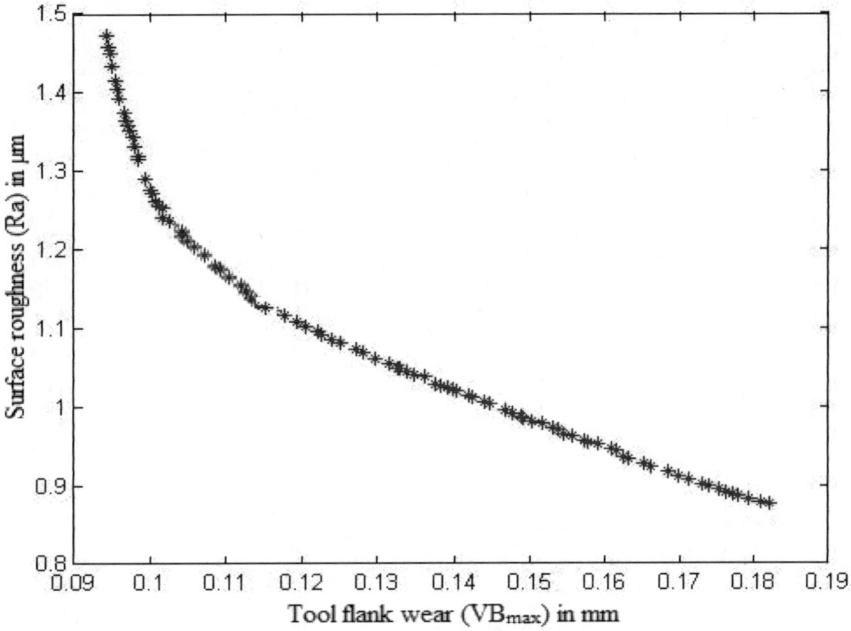

Figure 7. Optimal chart obtain through NSGA-II

From the experimental results presented in Table 2, the parameters listed in the experiment number 1 leads to minimum VB_{max} of 0.099 mm and corresponding Ra of 1.58 µm, where the cutting speed, feed rate, depth of cut and machining time are 50 m/min, 0.15 mm/rev, 1 mm and 4 min respectively. By optimizing NSGA-II, the same experimental value has been selected from the Table 5, serial number 3. The VB_{max} value is 0.099 mm and corresponding Ra is 1.23 µm and the pertinent parameters are cutting speed, feed rate, depth of cut and machining time are 50 m/min, 0.06 mm/rev, 0.5 mm and 2 min respectively. This indicates that values obtained from the optimization technique are in close agreement with the experimental values and more or less the same parameter settings. With reference to the minimum VB_{max} obtained through experimental results there are 7 set of readings with lesser VB_{max} obtained through NSGA-II are shown in Table 6.

Table 5 Optimal combination of parameter by NSGA-II

S.		Actual cutting parameters				Response variable	
No	Run	*V*	*f*	*d*	*t*	Y_1 (VB_{max}, mm)	Y_2 (*Ra*, μm)
1	1	50	0.13	0.5	2	0.094	1.47
2	2	150	0.05	0.6	2	0.182	0.86
3	3	50	0.06	0.5	2	0.099	1.23
4	4	50	0.10	0.5	2	0.095	1.37
5	6	84	0.05	0.6	2	0.130	1.05
6	10	150	0.05	0.7	2	0.159	0.94
7	11	121	0.05	0.6	2	0.166	0.91
8	16	50	0.11	0.5	2	0.094	1.41
9	24	75	0.05	0.6	2	0.122	1.08
10	29	50	0.05	0.6	2	0.101	1.20
11	33	141	0.05	0.5	2	0.175	0.88
12	35	81	0.05	0.6	2	0.127	1.06
13	43	115	0.05	0.6	2	0.154	0.95
14	45	146	0.05	0.5	2	0.179	0.87
15	49	50	0.12	0.5	2	0.094	1.44
16	50	65	0.05	0.5	2	0.144	1.12
17	52	136	0.05	0.5	2	0.171	0.89
18	59	56	0.05	0.6	2	0.107	1.15
19	60	50	0.05	0.5	2	0.100	1.21
20	61	105	0.05	0.7	2	0.147	0.98
21	69	61	0.05	0.6	2	0.111	1.14
22	72	125	0.05	0.6	2	0.163	0.92
23	73	100	0.05	0.7	2	0.142	1.00
24	75	50	0.07	0.6	2	0.098	1.26
25	76	91	0.05	0.6	2	0.135	1.03
26	79	110	0.05	0.7	2	0.150	0.97
27	83	50	0.09	0.5	2	0.096	1.33
28	84	71	0.05	0.6	2	0.119	1.09
29	90	50	0.08	0.6	2	0.097	1.29
30	93	95	0.05	0.6	2	0.138	1.01

The confirmation of the test outcome under the chosen optimum setting for the cases of VB_{max} and *Ra* are given in the Table 7. The predicted cutting performance is compared with the real cutting performance and an excellent conformity is obtained among their performances. Analyzing validation test result of Table 7, noticed that the calculated error is small. The error between the experimental and the predicted values for VB_{max} and *Ra* lie down within 1% and 2% respectively. Observably, this confirms outstanding reproducibility of the experimental conclusions.

Table 6 Optimal parameter for machining of Al/20%SiC using PCD tool

S. No	Run	Actual factors V	f	d	t	Response variable Y_1 (VB_{max}, mm)	Y_2 (Ra, μm)
1	1	50	0.13	0.5	2	0.094	1.47
3	4	50	0.10	0.5	2	0.095	1.37
4	16	50	0.11	0.5	2	0.094	1.41
5	49	50	0.12	0.5	2	0.094	1.44
6	75	50	0.07	0.6	2	0.098	1.26
7	83	50	0.09	0.5	2	0.096	1.33
8	90	50	0.08	0.6	2	0.097	1.29

Table 7 Validation test outcome for Al/20%SiCp MMC using PCD tool

Sl. No	Cutting speed (V) m/min	Feed rate (f) mm/rev	Depth of cut (d) mm	Machining time (t) min	VB_{max} (mm) Predicted	Actual	% Error	Ra (μm) Predicted	Actual	% Error
1	50	0.06	0.5	2	0.099	0.098	1	1.23	1.20	2

7 Conclusion

The machining parameter of particulate aluminum metal matrix composite using tipped PCD tool have been optimized via by means of Non-Dominated Sorting Genetic Algorithm (NSGA-II), and a non dominated solution set is obtained. The mathematical models developed for tool flank wear and surface roughness have been used for optimization. The choice of one solution over the other depends on the requirement of the process engineer. If the requirements of minimum tool flank wear or surface roughness, an appropriate mixture of cutting variables can be chosen. This technique will assist to enhance manufacture rate significantly by reducing machining time.

References

[1] Mannaa A and Bhattacharya B 2003 J Mater Process Technol. 140: 711–716

[2] M R Jadhav and U A Dabade 2016 IOP Conf. Ser.: Mater. Sci. Eng. 114 012122

[3] Diptikanta Das, Anil Kumar Chaubey, Bijaya Bijeta Nayak, Purna Chandra Mishra and Chandrika Samal 2018 IOP Conf. Ser.: Mater. Sci. Eng. 377 012110

[4] Seeman M, Ganesan G, Karthikeyan R and Velayudham 2010 Int J Adv Manuf Technol 48:613-624

[5] Vaccari J A and Lane C T 1993 American Machinist 11: 56-60

[6] El-Gallab M and Sklad M 1998 J Mater Process Technol. 83: 151–158, 277–285

[7] El-Gallab M and Sklad M 2000 J Mater Process Technol. 101: 10-20

[8] Lin J T, Bhattacharyya D and Lane C 1995 Wear 181–183: 883–888

[9] Andrewes C, Feng H and Lau W 2000 J Mater Process Technol. 102: 25–29

[10] Paulo Davim J 2002 J Mater Process Technol. 128: 100–105

[11] Kanagarajan D, Karthikeyan R, Palanikumar P and Paulo Davim J 2008 Int J Adv Manuf Technol. 2008, 36, 1124-1132

[12] Kuriakose S and Shunmugam M S 2005 J Mater Process Technol.170: 133–141

[13] Senthilkumar C, Ganesan G, and Karthikeyan R 2009 Proc. IMechE Vol. 224: 1399- 1407 Part B: J. Engineering Manufacture

[14] Deb K, Pratap A, Agarwal S and Meyarivan T. A 2002 IEEE Trans. Evol. Comput. 6(2): 182–197.

[15] Senthilkumar C, Ganesan G, and Karthikeyan R 2011 Trans. Nonferrous Met. Soc. China 21 2294–2300

[16] Seshadri A 2007 //http: //www.mathworks.com

Effect of Process Parameters in Friction Stir Welding of Dissimilar Aluminium Alloys

D.Muruganandam[a*], D.Raguraman[b] and L.A.Kumaraswamidhas[c]
[a,b]Jeppiaar Institute of Technology, Sriperumpudur, Kanchipuram, Chennai.
[c]Indian Institute of Technology Dhanbad, Dhanbad, Jharkand
Corresponding author: murudurai@gmail.com

Abstract. Friction stir welding was invented at Welding institute, United Kingdom in the year 1991 and has ever since been proved to be one of the best solid state joining methods for materials such as aluminium and magnesium. Some of the aluminium alloys which are not able to weld (Al-Cu, Al-Zn-Mg alloy) by fusion welding techniques, which produce defects and reduce the mechanical properties on the weld nugget could be welded using friction stir welding (FSW) successfully with excellent joint efficiencies. However effect of the process parameters on the properties of weldment have not been investigated fully. In this study, 5mm thick plates of aluminium alloys AA2024 (4.4%Cu, 0.6%Mn, 1.6%Mg) and AA7075 (2.5%Mg, 1.4%Cu, 0.2%Cr, 5.6%Zn) is selected for investigation. The FSW welding process was conducted on varying the welding process parameters such as tool rotation speed (RPM), welding speed (mm/min) and tool pin profile by keeping downward force (kN) constant. The properties such as defects, macro and microstructure, micro hardness, tensile and bend behavior on welded plates were studied. Sixteen sets of butt welds were produced using straight and tapered cylindrical with left hand threaded FSW tool pins. It has been found out that defect free weld could be produced with the help of tapered cylindrical with left hand threaded tool pin with tool rotation speed of 600 RPM, welding speed of 30 mm/min and downward force of 2.5 kN

1. Introduction

Friction stir welding involves an extensive utilization in aerospace, automobile, shipbuilding, and various production activities. The process shows the domination in non-heat treatable aluminum alloys over the fusion welding which is not feasible [1]. Thus basic survey both on the weld structure and on the relationship between micro structure with mechanical characteristics and process parameters have been initiated. A considerable influence of Friction Stir Welding is that the higher feasibility of welding dis-similar alloys compared to inconveniences with fusion welding procedures. One of the potential functions concerns the welding of strong performance materials, such as particle reinforced aluminium alloy, to wider structures formed from a poorer performance. A uniquely developed tool, formed from material that have a strong and wear resistant with respect to the material being joined, is twisted and plunged into the adjoining edges of the aluminium alloy to be welded. After an access of the tool pin to approximately the depth of the plate of aluminium alloy, the shoulder of the tool will just start penetrate into the aluminium plate due to the rotation of a tool. The rotating pin generates frictional heat on the aluminium alloy and causes the plastic deformation and movement of material flow from the advancing side of the tool to the retreating side where it relaxes and develops to a strong weld. [2-14].

2. Experimental Work

The friction stir welding of AA 2024 and AA7075 were preceded with butt joint configuration on

FSW machine which is displayed in Figure 2.1. The machine specifications are

1. Motor capacity : 12kW / 440V AC drive.
2. Spindle speed : 200 to 2000 RPM.
3. Downward Force : 0.5 to 10 kN.
4. Traverse feed : 10 to 150 mm/min.

Figure 1. Friction stir welding machine

2.1. Welding parameters

In this investigation, aluminium alloys AA2024 and AA7075 welding criteria such as transverse speed and rotation were altered and the downward axial force is maintained constant. The welding criteria are indicated in Table 2.1.

Table 2.1 Weld parameters – Friction stir weld of Dissimilar aluminium alloys

S.No.	Weld parameters	Values
1	Downward Force (kN)	2.5
2	Welding Speed (mm/min)	20,30,40,50,60
3	Tool Rotation Speed (RPM)	600,700,800,900

Tool parameters

The FSW tool pin were formed by **M2** HSS and which was hardened and tempered to 50 HRC. Three different pin profiles (i) straight pin (cylindrical with left handed threading) and (ii) taper pin (cylindrical with left handed threading) were used and are shown in Figure 2.2 (a-b).
The chemical composition for the tool pin is given in Table 2.2. Tool dimensions for the three pin profiles are given in Table 2.3.

IOP Conf. Series: Materials Science and Engineering **574** (2019) 012009 doi:10.1088/1757-899X/574/1/012009

Table 2.2 Tool material composition

Material/Alloy composition	C	Cr	W	Mo	V	Fe
M2	0.85	4.0	6.0	5.0	2.0	remaining

(a) Straight Cylindrical Threaded (b) Taper Cylindrical Threaded
Figure 2.2 Tool Pin profiles

Table 2.3 Tool Pin Profile Dimensions

Total length (mm)	Shoulder length(mm)	Shoulder dia (mm)	Pin length (mm)	Pin dia (mm)	Pitch (mm)	Taper pin dia (mm)
100	10	18	4.5	6	0.5	Major : 6 Minor : 5

Sixteen plates of butt-configuration weld were carried out for dissimilar AA2024 and AA7075 aluminum alloys; by altering the weld parameters tool transverse speed, rotational speed and constant downward force with the utilization of straight pin and taper pin. Eight sets of welds were made with the help of straight tool pin (ST1 to ST8) and another eight sets with taper tool pin (TT1 to TT8). The welding process parameter window is given in Table 2.4.

Table 2.4 Welding Parameters for dissimilar welding of AA2024-AA7075

Specimen Code No.	Tool (Cylindrical shaped and threaded)	Tool Rotation Speed (RPM)	Welding Speed (mm/min)	Downward Force (kN)
ST1	Straight Tool	600	30	2.5

ST2	Straight Tool	700	30	2.5
ST3	Straight Tool	800	30	2.5
ST4	Straight Tool	900	30	2.5
ST5	Straight Tool	700	20	2.5
ST6	Straight Tool	700	40	2.5
ST7	Straight Tool	700	50	2.5
ST8	Straight Tool	700	60	2.5
TT1	Taper Tool	600	30	2.5
TT2	Taper Tool	700	30	2.5
TT3	Taper Tool	800	30	2.5
TT4	Taper Tool	900	30	2.5
TT5	Taper Tool	800	20	2.5
TT6	Taper Tool	800	40	2.5
TT7	Taper Tool	800	50	2.5
TT8	Taper Tool	800	60	2.5

3. Result and Discussion

TT1 to TT8 welds were carried out with the taper cylindrical threaded tool pin. Only tool rotation speed is altered in the welds TT1 to TT4 and remaining parameters were maintained constant. In TT5 to TT8 welds, welding traverse speed is altered and other parameters were maintained constant which is presented in Table 2.4. Weld plate TT5 got the worm hole defect in visual inspection which is shown in Figure 3.1, on the other plates no such irregularities were discovered. TT1 weld plate revealed a smooth appearance at the weldment whereas poor surface condition like sand paper display was noted at the weldment of the plates TT2 to TT8; this is due to the tremendous heat caused during welding [2].

Fig. 3.1 Worm hole defect

IOP Conf. Series: Materials Science and Engineering **574** (2019) 012009 doi:10.1088/1757-899X/574/1/012009

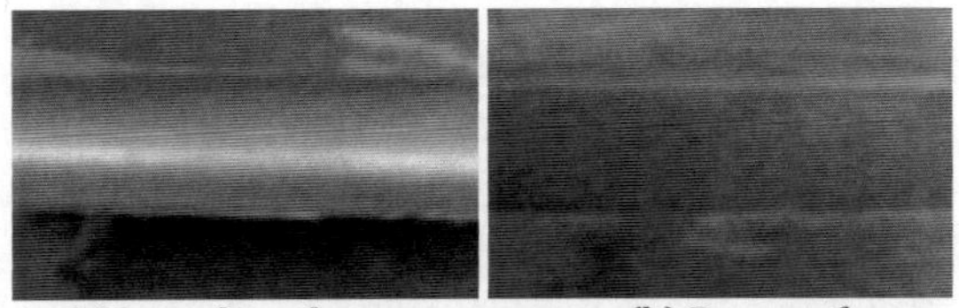

(a)Smooth surface (b) Poor surface

Fig. 3.2 Surface Qualities at the Weldment

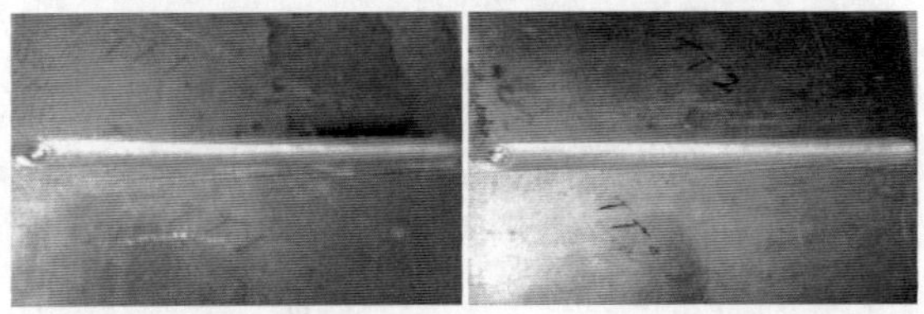

(a) TT1 Weld Plate *(b) TT2 Weld Plate*

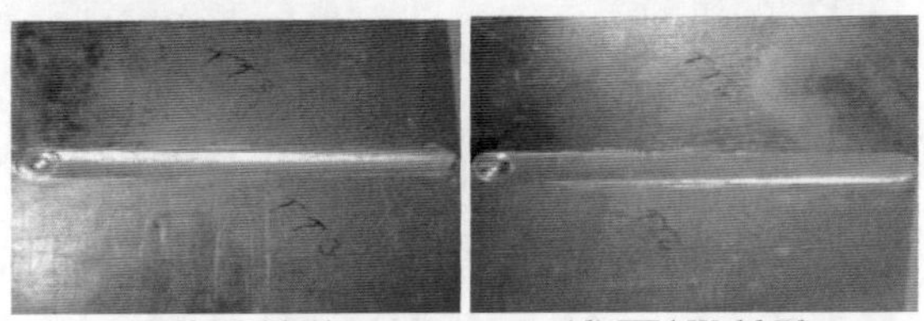

(c) TT3 Weld Plate *(d) TT4 Weld Plate*

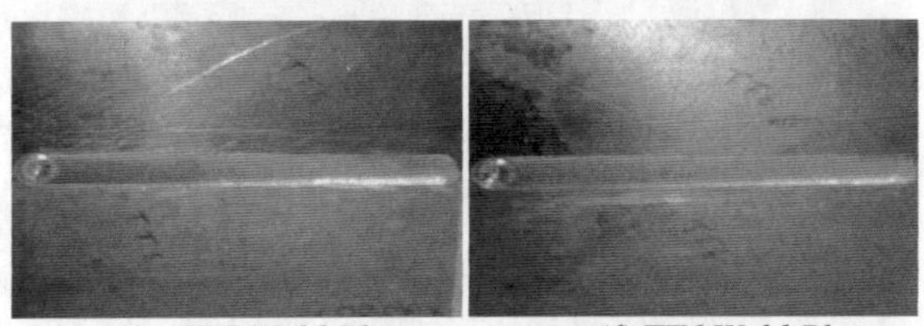

(e) TT5 Weld Plate *(f) TT6 Weld Plate*

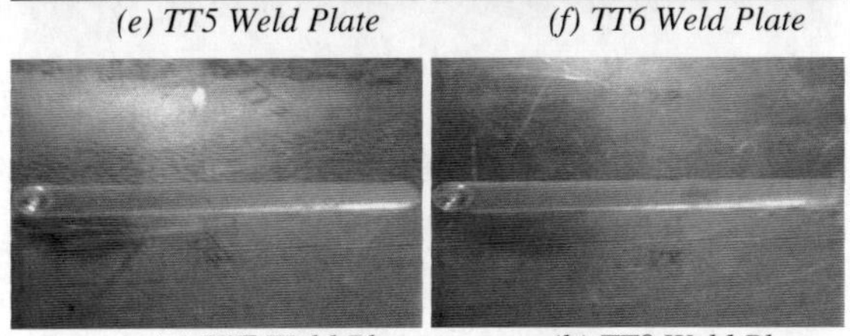

(g) TT7 Weld Plate *(h) TT8 Weld Plate*

Fig. 3.3 TT Series Butt Welded Plates

The particles of aluminium contribute to hold themselves to the surface of the shoulder of the tool and are snatched out of the weld face surface and turned over to another position as shown in Figure 3.2. The welded plates are displayed in Figure 3.3.

ST1 to ST8 welds were made with the straight cylindrical threaded tool pin. Similar to TT series welded plates, particularly tool rotation speed is altered in the welds ST1 to ST4 by holding other parameters as constant. In ST5 to ST8 welds, welding traverse speed is varied and other parameters were kept constant which are given in Table 2.4. Weld plates ST3, ST4, ST5 and ST8 got the worm hole defect during visual analysis, on the other plates no such defects was observed. Like TT series welds poor surface appearance (like sand paper surface) was detected at all the weldment of ST1 to ST8, which is revealed in Figure 3.2 and the welded plates are shown in Figure 3.4.

IOP Conf. Series: Materials Science and Engineering **574** (2019) 012009 doi:10.1088/1757-899X/574/1/012009

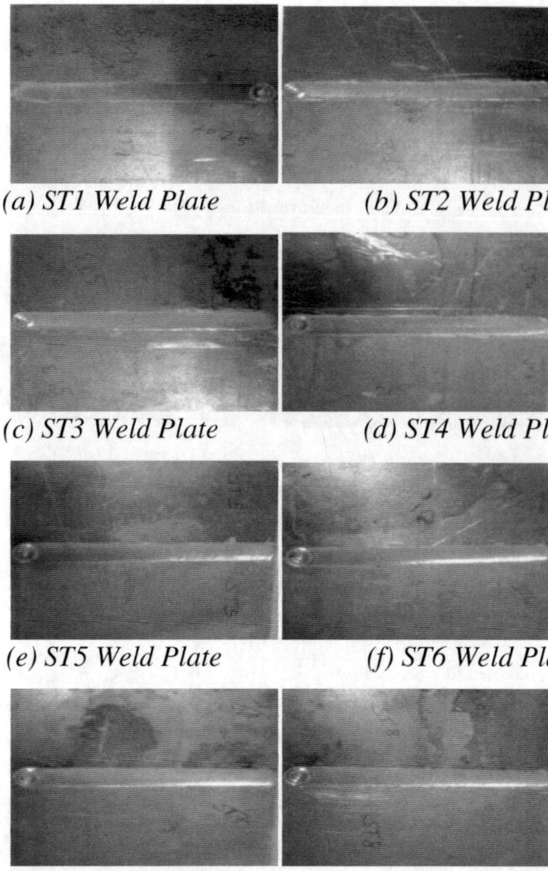

(a) ST1 Weld Plate *(b) ST2 Weld Plate*

(c) ST3 Weld Plate *(d) ST4 Weld Plate*

(e) ST5 Weld Plate *(f) ST6 Weld Plate*

(g) ST7 Weld Plate *(h) ST8 Weld Plate*

Fig. 3.4 ST Series Butt Welded Plates

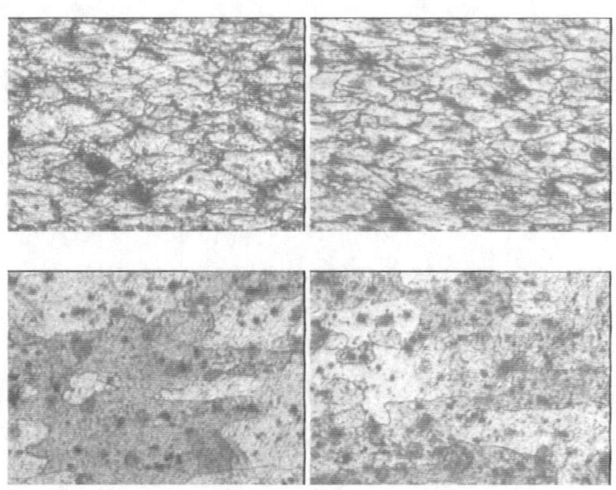

(a) AA 2024 Aluminum Alloy *(b) AA 7075 Aluminum Alloy*

Fig. 3.5 Base metal microstructure

Microstructure:

The microstructure of the welded joint is properly arranged in to four sectors namely base metal (BM), thermo mechanically affected zone (TMAZ), heat affected zone (HAZ) and weld nugget zone (WNZ). The microstructure of the base alloys is presented in Figure 3.5 and microstructures of different region of the weld region for dissimilar aluminium alloy AA2024-AA7075 are shown in Figure 3.6.

The microstructure of AA2024 indicated the presence of elongated grains of mean size of 10μm, whereas AA7075 showed the presence of large grains of mean size of about 40 μm.

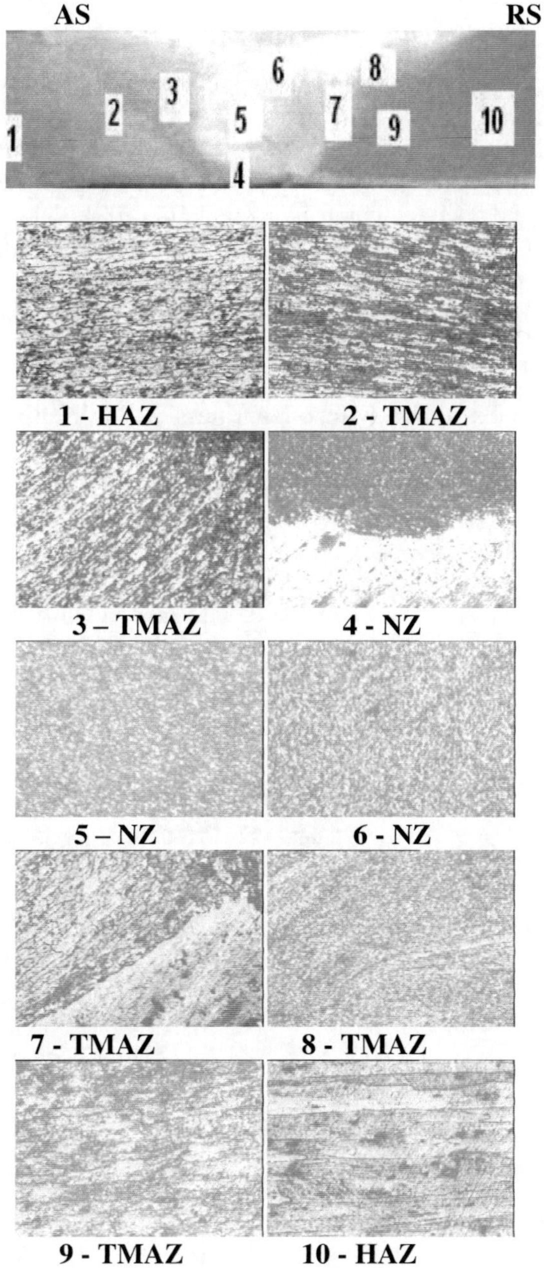

Fig. 3.6 Microstructure of Different Region at the weldment

The microstructure of the various zones of the TT1 welded plate is presented in Figure 3.6. It indicates the microstructure at the weld nugget zone is comprised of the fine equiaxed recrystallized grains related to the microstructure of the base metals.

The initial grain and sub grain boundaries are transformed into nearly fine equiaxed recrystalization due to high temperature and high rate of deformation in this region. Whereas in thermo mechanically affected zone and heat affected zone microstructure gets larger, elongated grains boundaries in the both advancing and retreating side.

Microhardness:

The micro hardness test was conducted for TT1, TT2, ST1 and ST2 welded plates. The tests were conducted for every 2 mm gap from the weld center point (on weld nugget). The results are tabulated in Table 3.1 and are shown in Figure 3.7. The hardness values of heat affected zone (HAZ), thermo mechanically affected zone (TMAZ) and the weld nugget (WN) is less than that of base metal (BM). But the hardness in the weld nugget region is considerably increased in correlation to the HAZ and TMAZ which is accountable to the grain refinement in WN, due to intensive stirring which is shown in Figure 3.7. In frequent situations the hardness of the advancing side was observed to be higher than the hardness of the retreating side. This could be due to the initial high hardness of AA7075 (160 VHN) in comparison with the hardness of AA2024 (154 VHN).

Table 3.1. Micro Hardness at the Weldment

Weld	Retreating side			Centre point			Advancing side		
Dist. (micro meter)	**-8**	**-6**	**-4**	**-2**	**0**	**2**	**4**	**6**	**8**
TT1	122	126	139	149	152	151	147	135	133
TT2	123	130	137	148	153	150	145	131	130
ST1	121	128	135	147	152	150	148	130	125
ST2	120	125	136	148	151	149	141	130	122

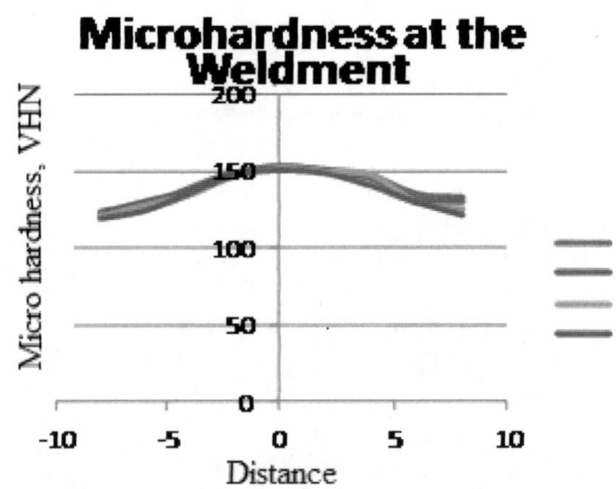

Fig. 3.7. Micro hardness at the weldment

Tensile Properties:

For AA2024-AA7075 alloy, TT1, TT2, ST1 and ST2 welded plates were tested and the results are given in Table 3.2. The tested specimens are shown in Figure 3.8. Compared to base metals, the welded plate TT1 tensile properties were considerably decreased. But in other plates TT2, ST1 and ST2 tensile properties were largely decreased due to the effect of the tunnel defect in the weldment. The fractured surface of the tensile test specimens are displayed in Figure 3.9. The location of the fracture is between the weld nugget zone (NZ) and thermo mechanically affected zone (TMAZ). The fracture surface shows a groove like feature at the bottom of the tunnel defect, similar feature in the fracture surface of the FSW AA6061 and AA2219[4,9].

(a) TT1 Specimen

(b) TT2 Specimen

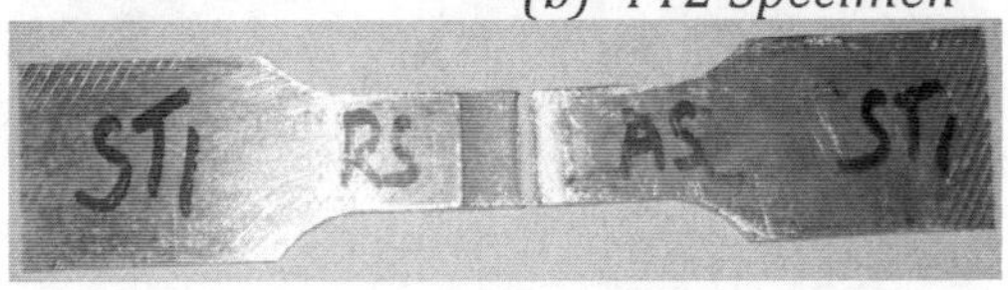

(c) ST1 Specimen

(d) ST2 Specimen

Fig. 3.8 AA2024-AA7075 Tensile Tested Weld Specimen

TT1 Tensile Specimen

TT2 Tensile Specimen

ST1 Tensile Specimen

ST2 Tensile Specimen

Fig. 3.9. Fractured Surface of Tensile Specimen at weldment

IOP Conf. Series: Materials Science and Engineering **574** (2019) 012009 doi:10.1088/1757-899X/574/1/012009

Table 3.2 AA2024-AA7075 Tensile Test Results

Combination	0.2% proof stress (MPa)	Tensile Strength (MPa)	Elongation (%)	Joint efficiency (%)
AA2024 **(Base material)**	348	459	20	
AA7075 **(Base material)**	384	439	17	
TT1	319	335	15.5	74.6
TT2	262	278	11	61.9
ST1	113	119	9.2	26.5
ST2	109	117	6	26

Bend behavior:

For AA2024-AA7075 alloy, only TT1 welded plate welded plates were tested for both face and root bend and the results are given in Table 3.3. The tested specimens are displayed in Figure 3.10.

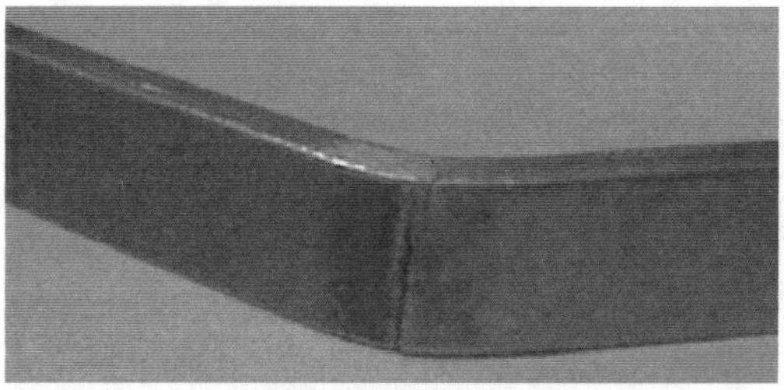

(a) Root bending

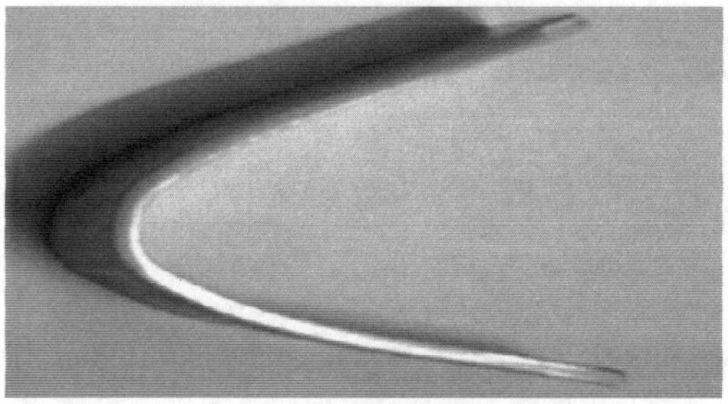

(a) Face bending

Fig. 3.10. AA2024-AA7075TT1 Root and Face Bend Tested Weld Specimens

The samples reveal that, face bend of the welded plate allows very high bend angles passes the bend test. In weld nugget no cracks were observed. But in the root bend the cracks were found after 42 degrees of the bend test.

Table 3.3 AA2024-AA7075 Bend Test Results

Combination	Root bend	Face bend
TT1	Cracks observed after 42 degrees of bend	No cracks observed

4. Conclusion

- Quality welds could be produced with the taper cylindrical threaded tool pin, at 600 RPM tool rotation speed, 30 mm/min welding speed and 2.5 kN downward force. No defect occurred in weld nugget region.
- The hardness is increased in the weld nugget zone compared to the other zones due to the fine and equiaxed grains observed in the weld nugget region. But hardness values of WNZ, TMAZ and HAZ is decreased compared to the base metal.
- Tensile strength of the welded plate TT1 decreased to 335 MPa compared to the base metal value of about 450 MPa, but in other welded plates properties were largely decreased.
- The welded specimens passed only face bend test allowing for very high bend angles and no cracks were observed in weld nugget. But cracks were observed after 42 degrees of bend during the root bend test.

References

[1] G. O. Young, “Synthetic structure of industrial plastics (Book style with paper title and editor),” in *Plastics*, 2nd ed. vol. 3, J. Peters, Ed. New York: McGraw-Hill, 1964, pp. 15–64.
[2] Aluminium Alloy Welding Book, Mishra R.K (2006), Vol.2.
[3] Adamowski J. (2007), ‘Analysis of FSW welds made of aluminium alloy AW6082-T6’,Archives of Materials Science and Engineering ,Volume 28, Pages 453 - 460.
[4] Bahemmat P (2008), ‘Experimental study on the effect of rotational speed and tool pin profile on AA2024 aluminium friction stir welded butt joints’, In: Proceedings of ECTC ASME early career technicalconference; 2008. Miami, Florida, USA.
[5] Balasubramanian V (2008), ‘Influences of tool pin profile and tool shoulder diameter on the formation of friction stir processing zone in AA6061 aluminium alloy’, Materials and Design 29, page 362–373.
[6] Cavaliere P (2005), ‘Mechanical and micro structural behaviors of 2024–7075 aluminium alloy sheets joined by friction stir welding’, International Journal of Machine Tools & Manufacture, Vol 46(6), Pages 588-594
[7] Chen Z. W. (2007), ‘Tool-workpiece interaction and shear layer flow during friction stir welding of aluminium alloys’. Trans. Non ferrous mater. Soc. China (17), Pages s258 - s261.
[8] Chunlin Dong (2009), ‘Microstructure and mechanical properties of friction stir welded joints in 2219-T6 aluminum alloy’, Materials and Design 30, page 3460 – 3467.
[9] Chionopoulos S.K. (2007), ‘Effect of tool pin and welding parameters on friction stir

welded (fsw) marine aluminium alloys', Conference: 3rd International Conference on Manufacturing Engineering (ICMEN), At Chalkidiki, Greece.
[10] Elangovan K (2008), 'Influences of tool pin profile and welding speed on the formation of friction stir processing zone in AA2219 aluminium alloy', Journal of materials processing technology 200, page 163–175.
[11] McClure J.C. (2007), 'Effect of Pin tool Shape on Metal Flow During Friction Stir Welding'.
[12] Nandan R. (2007), 'Recent Advances in Friction Stir Welding – Process,Weldment Structure and Properties', progress in material science 53 (6), pages 980 – 1023.
[13] Padmanaban G (2009), 'Selection of FSW tool pin profile, shoulder diameter and material for joining AZ31B magnesium alloy – An experimental approach', Materials and Design 30, page 2647–2656.
[14] Scialpi A (2007), 'Influence of shoulder geometry on microstructure and mechanical properties of friction stir welded 6082 aluminium alloy', Materials and Design 28, page 1124–1129.
[15] Zadpoor A A, Sinke J (2007), 'The effects of friction stir welding on the mechanical properties and microstructure of 7000 series aluminium tailor-welded blanks', International journal of material forming(1), pages 1311 – 1314.

Effect of Nano Filler Reinforcement on MechanicalProperties of Epoxy Composites

Suresha B[1 *], C A Varun[1], Indushekhara N M[1], Vishwanath H R[1] , Venkatesh[1]
[1] Department of Mechanical Engineering, The National Institute of Engineering, Mysuru, India

E-mail: sureshab@nie.ac.in

Abstract

Nano-sized functional fillers, such as Nano-clay (N-C) and Nano-calcium carbonate (N-CC), have attracted a great deal of awareness due to their unexpected intrinsic properties. Extensive research has been done on clay material for epoxy Nanocomposites but only a few have ventured into a comparison with other Nanofillers. In this paper, a study has been made of the mechanical properties of epoxy containing N-C and N-CC fillers. These composites were studied for their mechanical properties such as hardness, tensile, flexural and impact strength as per ASTM standards. Fraction of void cohort in the samples was quantified following the rule of mixture. Experimental results demonstrated that incorporation of N-C fillers expanded the hardness, impact strength and module of epoxy Nanocomposites. However, the tensile and flexural properties have shown declining trend for N-C/N-CC Nanocomposites. The morphological and fractographic features were examined utilizing scanning electron microscopy.

1. Introduction

Over the past few decades, engineering polymers have replaced many of the conventional materials in various applications [1,2]. This is feasible because of the benefits that polymers offer over conventional materials. In most of the engineering applications, the properties of polymers are modified using micro/Nanofillers and fibers to suit the high modulus /high strength requirements [3-5]. The principal driving forces that led to the interest and investment in polymer composite materials in the mid of 19th century were the designer's demand for lower weight and higher rigidity for aero or space structures, electronics, sporting goods, and building structures [6-8]. In two-phase polymer matrix composites, numerous inorganic fillers with micron size inorganic particles, for example calcium carbonate, glass beads and talc have been utilized widely to upgrade the mechanical properties of polymers. Such properties can surely be custom-made by changing the volume portion, shape, and size of the filler [9,10]. The mechanical properties of the composites reinforced with micron sized functional fillers are inferior to those loaded with Nano-sized particles of a similar material. The progress of multifunctional composite materials possessing novel properties has been achieved with the incorporation of Nanofillers which has overcome numerous disadvantages of micro filler reinforced composites. Thus, such novel polymeric matrices viz., thermoset and thermoplastics reinforced with fillers having Nano-dimensions usually less than 50 nm, which are especially termed as polymer Nanocomposites (PNC's), have attracted exceptional research interest [11-14]. Through the variation in filler dimension from micrometer to Nanometer scale, the surface area to volume ratio has been found to alter by three orders of magnitude leading to the drastic changes in morphological features as well as in their mechanical properties in polymeric matrices [15-17]. The most significant component that influences the interfacial bond is accepted to be the mechanical loading, chemical interaction and physico-mechanical boundary layers. Chemical modification of the filler surface to improve bond has turned out to be progressively significant. From the earlier reported works, it can be concluded that to improve the filler/matrix adhesion and filler scattering surface modification of the filler is very important. Interfacial bond quality relies upon the characteristic wetting capacity of the polymer and viability of the coupling agents. Consolidation of different added substances or coupling agents in these systems helps to promote the adhesion the matrix/filler interface and, thus, load transfer between the filler and the matrix improves the degree of filler scattering, builds the filler loading in the composite and improves the processability and mound ability. Several research groups are active in this area of research and in the most recent decade various Nano-sized fillers were presented [11-17]. Additionally, a few researchers have made endeavors to improve the mechanical and different properties of epoxy matrix by modifying Nano-calcium silicate, Nano-alumina, Nano-SiO2, Nano-TiO2, and Nano-CaCO3 [18, 19]. In the present work, the impact of Nano-clay and Nano-calcium carbonate on the mechanical properties of the composites of pristine epoxy has been researched. As far as anyone is concerned there are no genuine examinations in regards with the impact of Nano-clay and Nano-calcium carbonate on the mechanical properties of pristine epoxy yet. Hence, the purpose of this experimental work was to evaluate the effect of Nanofillers namely Nano-clay (N-C) and Nano-calcium carbonate (N-CC) on the physico-mechanical properties of epoxy Nanocomposites polymerized with conventional light-curing and additional post-curing.

2. Materials, fabrication and testing

2.1 Materials

The thermoset resin epoxy (Lapox 12) and the hardener (K6) were provided by Atul Limited, Gujarat, India. Nano-clay (N-C, size 20-30 nm) and Nano-calcium carbonate (N-CC, 20-40 nm) were purchased from Global Nano-Tech Bangalore, India. The N-C, N-CC and the epoxy resin possess Young's modulus of 178 GPa, 69.9 GPa, and 3.35 GPa respectively and a density of 2.35 g/cm^3, 2.71 g/cm^3, and 1.15 g/cm^3 respectively.

2.2. Fabrication of composites

A pictorial representation of the steps involved in the fabrication of the composites has been illustrated in Fig. 1.

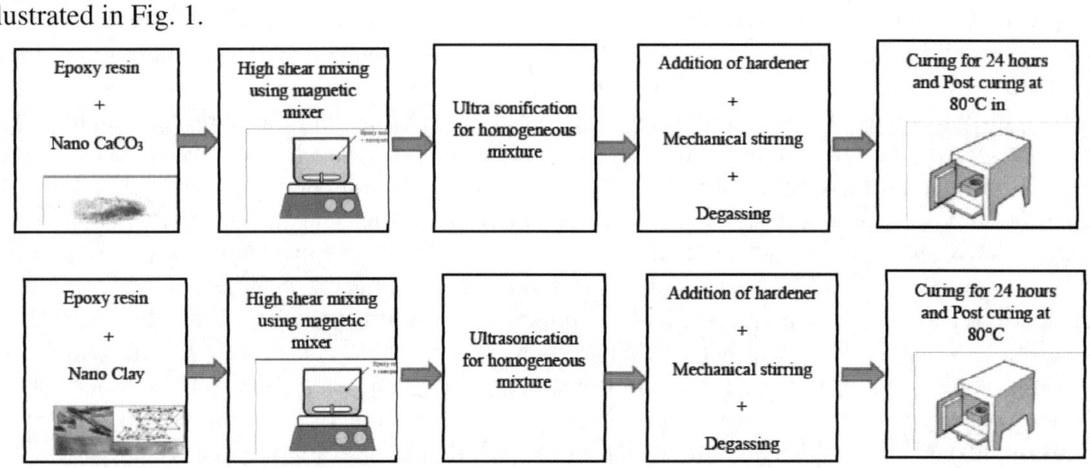

Figure 1 Fabrication of N-C/N-CC epoxy Nanocomposites

Epoxy resin was preheated to 60 ^{0}C before desired amount of Nano clay (N-C) / Nano calcium carbonate (N-CC) was introduced and mixed using mechanical stirrer for 20 min. To reduce the viscosity of resin/filler mix and facilitate uniform mixing, temperature was maintained at 60 ^{0}C for the whole duration of mixing using a hot plate. Further, to remove entrapped bubbles from the mixture, degassing was done for 30 min. Subsequently this mixture was ultra-sonicated for 40

IOP Conf. Series: Materials Science and Engineering **574** (2019) 012010 doi:10.1088/1757-899X/574/1/012010

min. Curing agent (K6) was added to the resin/filler mixture at 100:10 ratio and carefully hand mixed to avoid introduction of any air bubble. After, it was casted using open aluminummoldswith dimensions 200 mm × 150 mm × 5 mm. The curing was done at room temperature for 24 h followed by post curing at 85 ^{0}C for 6 h. The weight percentages of N-C/N-CC were varied from 0, 1, 3 and 5 to investigate the role of Nanofillers on mechanical properties of epoxy composites.

Seven different formulations were fabricated as per the compositions summarized in Table 1. For test coupons 1N-C to 5N-C and 1N-CC to 5N-CC, the filler loading varied from 1, 3 and 5 wt. %.

2.3Testing of epoxy Nanocomposites

2.3.1. Density test

The specimens were tested according to ASTM standards. The density of the specimens was measured in accordance with ASTM D792 with the dimension of 6mm×6mm. Equations 1 and 2 are used for calculations of theoretical density ($\boldsymbol{\rho_{Ct}}$) and void fraction ($\boldsymbol{V_v}$)of Nanocomposites fabricated.

$$\rho_{Ct} = \frac{1}{\frac{W_m}{\rho_m}+\frac{W_p}{\rho_p}} \tag{1}$$

where W and ρ represents weight fraction and density of the materials respectively. The suffixes $m$and p represent matrix and particulate respectively.

$$V_v = \frac{\rho_{ct}-\rho_{ca}}{\rho_{ct}} \tag{2}$$

where ρ_{ct} and ρ_{ca} represents theoretical and actual density of Nanocomposites respectively.

2.3.2 Hardness test

The hardness was measured on a Shore D hardness scale governed by ASTM D2240 with indentation of 0-2.5mm depth and 0.79 mm diameter by the application of 50 N load.

2.3.3 Tensile test

Tensile tests were carried out using universal testing machine (100 kN, Kalpak Instruments and Controls, Pune, India) at room temperature and constant crosshead speed of 10 mm/min. (Dog-bone shaped samples were made). Young's modulus, ultimate tensile stress and elongation at break of epoxy and epoxy Nanocomposites were measured. At least 6 composite samples were tested for each composition according to ASTM standard D638-08. Equation 1 and 2 are used to calculate tensile modulus.

2.3.4 Flexural test

The flexural test was also carried out on the same instrument in obedience with ASTM D790 with a cross head speed of 10mm/min. The length to thickness ratio was kept constant and the value was 16:1. The width of the specimen was 12mm. The below equations 3, 4 are used for calculation of slope and flexural modulus. Equations 3 and 4 are used for calculations of flexural strength (σ_f) and flexural modulus (E_f)of Nanocomposites fabricated.

$$\sigma_f = \frac{3FL}{2bd^2} \tag{3}$$

$$E_f = \frac{FL^3}{4ybd^3} \tag{4}$$

where F = peak load (N), L = length of span (mm), y = deflection at peak load (mm), b = width of specimen (mm) and d = depth of the specimen (mm).

2.3.5 Impact test

Izod Impact test was conducted on a notched specimen corresponding to ASTM D256 with R1 type hammer with range of 0-2.71J on a computerized Izod impact tester. The notch was made using a notch cutter and the width of the specimen was restricted to 10.16mm. The fracture surface of the neat epoxy and their Nanocomposites was analyzed by a JEOL scanning electron microscope (SEM).

3. Results and discussion

3.1. Measured Density

In general, the quality of composite slab fabricated, and surface hardness of a material has a great influence on its other mechanical properties. A good cast-slab of composite with hard surface of the material often results in improved mechanical properties. So, it was essential to study the influence of the Nanofillers namely, Nano-clay and Nano-calcium carbonate (N-C and N-CC), based onactual density and hardness on Epoxy. Mass density is a material property which is of prime significance in a few weight delicate applications. Density of a composite relies on the relative proportions of matrix and reinforcing materials. There is dependably a contrast between the measured and the theoretical density of a composite because of the presence of voids and pores. These voids fundamentally influence a portion of the mechanical properties and even the performance of composites. Higher void content typically means lower fatigue resistance, more prominent helplessness to water penetration and weathering. Table 1 shows the role of wt% of N-C and N-CC on the measured density of the Ep composites. The incorporation of N-C and N-CC increased the density of the Ep matrix material, verifying the usefulness of the Nanofillers as particulate reinforcement. It was found that the density of the epoxy-based composite has increased with increase in N-C or N-CC loading. The density of 5 wt% of N-C composite showed 1.74% enhancement and N-CC composite showed 1.82% enhancement in density with respect to neat epoxy. It is evident from Table 1 that the measured density and void percentage of epoxy increase with the incorporation of N-C and N-CC Nanoparticles.

Table 1 Measured, theoretical density and % void fraction

Composites	Measured density g/cm^3	Theoretical density g/cm^3	Void fraction %
Epoxy	1.152	1.15	0.17
1N-C	1.153	1.156	0.17
3N-C	1.163	1.168	0.43
5N-C	1.172	1.182	0.85
1N-CC	1.155	1.156	0.09
3N-CC	1.165	1.17	0.43
5N-CC	1.173	1.184	0.93

3.2 Shore-DHardness

This has happened due to formation of agglomerations due to mechanical mixing of higher weight fraction of fillers with the matrix. The agglomeration of fillers leads to the formation of in the fabrication of Nano clay/E-glass/epoxy composites also support the results [21].Fig. 2 shows the role of the wt% of N-C and N-CC on the hardness (Shore-D) of the Ep composites. The incorporation of Nanofiller increased the hardness of the Ep composites. The hardness of the N-C filled Ep composites increased gradually with the increase of filler loading. It has been found that there is an increase of 12.5 % of hardness when N-CC was incorporated and an increase of 14.9% when N-C was incorporated. This can be explained by the fact that N-C and new generation of processed clay consists of layered silicates when stacked together is a very hard and tough material. The agglomerated Nanoparticles have resisted the indentation which is a consequence of sudden load applied on it [23].

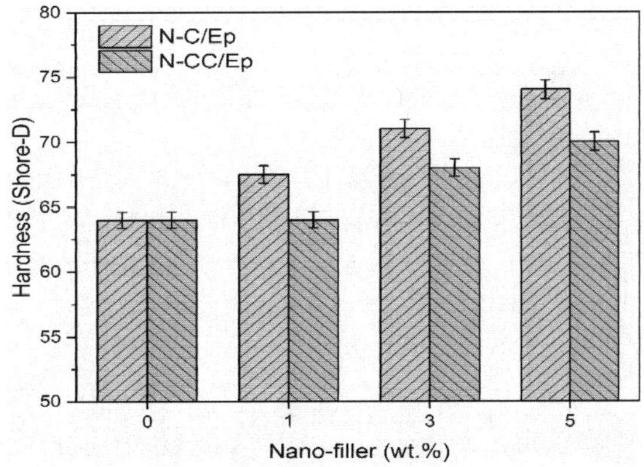

Figure 2 Hardness of N-C/N-CC epoxy Nanocomposites

3.3 Role of Nano fillers on tensile properties

It is evident from the plot (Fig. 3) that tensile strength, irrespective of the percentage of the fillers loaded, is lower than pristine Epoxy. N-C filler has caused a decrease in the strength by 22.2% for 5% N-C compared to neat epoxy. However, N-CC fillers have decreased the strength by 33.3% for 5% N-C composite.

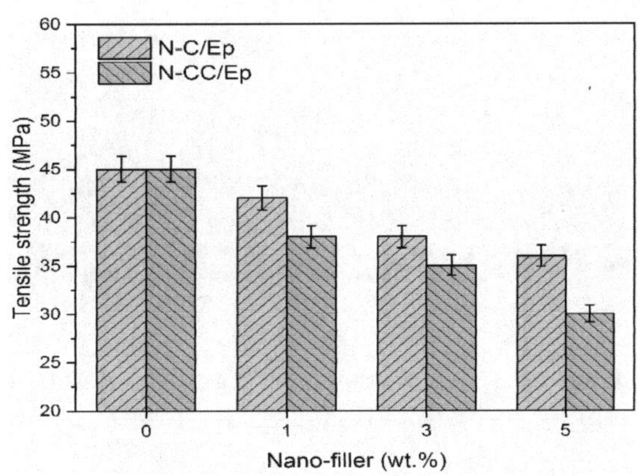

Figure 3 Tensile strength of N-C/N-CC epoxy Nanocomposites

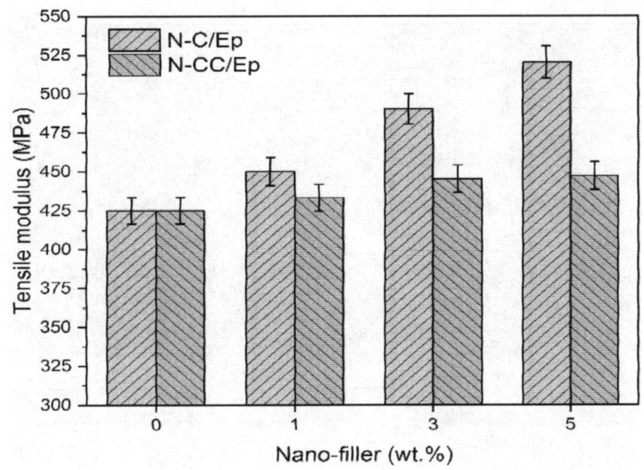

Figure 4 Tensile modulus of N-C/N-CC epoxy Nanocomposites

The serious local plastic deformation in matrix around the particles happens during the tension deformation process. The brittle nature of epoxy makes it very sensitive to the defect sites existing

in the samples. The agglomerated structure of Nanofiller particles would act as stress concentration site when the applied stress induces the interfacial failure between particles and matrix [24]. These defects generate stress concentration sites within the matrix, resulting in decreased tensile and flexural strength.
Fig. 4 shows the variation of tensile modulus with different weight fractions of NC and N-CC. It can be observed that tensile modulus of the matrix has increased up to 23.5% for 5% N-C reinforced composite and has been constant for N-CC reinforced composite. The addition of Nanoclay reduces the flexibility of the polymer chains, which leads to the increase in young's modulus.

3.3.1 Fractographic analysis of tensile failed specimens
According to the fractographic analysis, the rough surface and the torturous crack path on the N-C composite can be observed for both the materials. Fig. 5 shows the agglomerated particles on the crack paths of specimens.

Figure 5 SEM image of tensile failed N-CC specimen

The non-uniform distribution of particulates with agglomerations can be observed in the specimens. Fig. 6 shows the of pit formation of pit after the tensile failure of the specimen.

Figure 6 SEM image of tensile failed N-C specimen

3.4 Role of Nano fillers on flexural properties
Flexural strength and flexural modulus were determined for the post-cured N-C/N-CC epoxy composites as per ASTM D790 standard. In general, the flexural properties of the composites are

the combined effect of tensile, compression and shear properties. High elastic fillers usually improve the flexural properties of the composites. Fig. 7 and 8 show the flexural strength and modulus as a function of N-C/N-CC filler loading respectively. From the figures it can be seen that incorporation of N-C/N-CC particles leads to reduction in flexural strength of the composites. These properties were significantly affected for higher loading of Nanofillers (> 3 wt.%).Figure 8 shows the variation of flexural modulus with different weight fractions of NC or N-CCNanocomposites. It can be observed that there is about 6% increase in the flexural modulus of N-C composite and 17.64% decrease in flexural modulus of N-CC. Further, the effect of N-C is more pronounced in enhancing the flexural modulus of epoxy compared to N-CC filled epoxy Nanocomposites. The increase in flexural modulus of N-C/epoxy Nanocomposites may be due to intercalation of Nano clayin the epoxy matrix, uniform dispersion and good interfacial adhesion that resulted in improved flexural modulus.

3.4.1. Fractographic analysis offlexural failed specimens

The qualitative evaluation of the interface property could be obtained from the fractured surface of composite materials as depicted in Figs. 9 and 10.

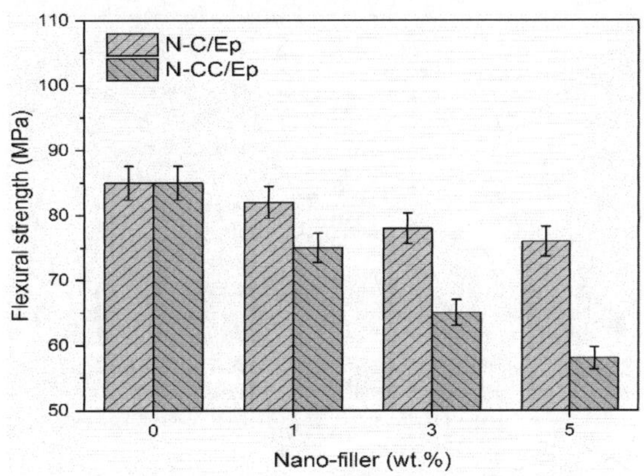

Figure 7 Flexural strength of N-C/N-CC epoxy Nanocomposites

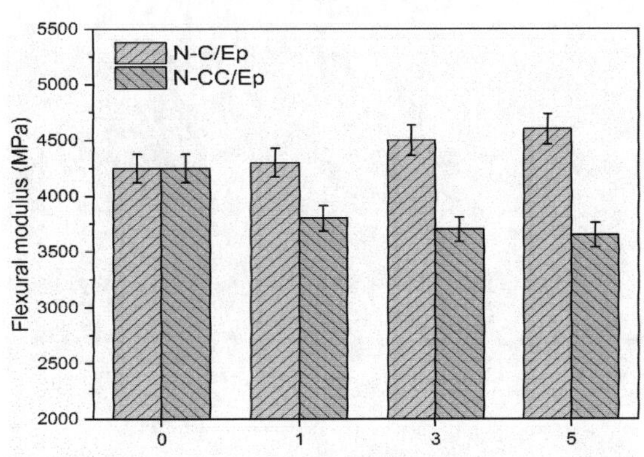

Figure 8 Flexural modulus of N-C/N-CC epoxy Nanocomposites

The SEM fractographs shown in Figs. 9 and 10 shows network of micro-cracks in the composites after the bending failure. This could be attributed to minor crack ignition ad propagation in the composites because of poor bonding of Nano particles with the epoxy matrix. The micro cracks in the composites can create a larger surface area that may lead to lower flexural strength of the composites. This is expected due to weak interfacial bonding between the filler and matrix. The

flexural modulus of composites may depend on the stiffness of high elastic modulus filler. The flexural modulus of composites is shown in Fig. 8. The flexural modulus of N-C/epoxy Nanocomposites improved by approximately 6% from 4250 MPa to 4500 MPa compared to neat epoxy. The results clearly show the effect of N-C filler enhancing the flexural modulus as depicted in fractograph in Fig. 10.
The morphology of fracture surface of N-CC flexural failed specimen has relatively less torturous paths when compared to N-C. This indicates that the resistance to crack propagation is less for N-CC and has led to greater reduction off strength of the matrix. N-C composite shows scratches and torturous paths on the fracture surface. These scratches are due to the particle peel-off from the material in which the entire Nano particle/agglomerated particles are removed from the matrix at the time of fracture. This leads to pitformation. The pit formation is observable in Fig. 10. During this process, the particles have scratched the surface and tried to resist the propagation of crack. This process has diminished the deteriorating effect of agglomerates on the flexural properties in N-C composites when compared to N-CC composites.

Figure 9 SEM image of Flexural Failed N-C Nanocomposites

Figure 10 SEM image of Flexural Failed N-CC Nanocomposites

3.5 Impact strength

The impact strength has considerably enhanced with increase Nano filler particles loading. Nano filler particles act as stress concentration sites, which promote cavitation at the particle-matrix boundaries during loading. The cavitation could release the plastic constraints and trigger mass plastic deformation of the matrix, leading to improved toughness. There is 62% increase in the impact energy for 5wt. % N-CC composite. There is an increase of 33.33% in impact energy for 3wt.%N-C composite. Impact strength has increased with increase in particle loading for N-CC whereas it has increased up to 3 wt. % and has decreased for further loading. The Table 2 shows the variation of impact energy with particle loading for N-CC and N-C composites.

Table 2 Impact strength of N-C/N-CC epoxy Nanocomposites

Composites	Impact strength (J/m)
Ep	2.8
1N-C	3.4
3N-C	4.1
5N-C	3.1
1N-CC	3.8
3N-CC	4.0
5N-CC	7.5

4. Conclusions

The mechanical properties of N-C and N-CC reinforced epoxy Nanocomposites was studied with respect to the filler loading. The following conclusions were drawn from the experimental results.

- The tensile strength and flexural strength of N-C and N-CC reinforced epoxy Nanocomposites dropped below that of neat epoxy.
- The tensile and flexural moduli of N-C reinforced epoxy Nanocomposites increased with increase in filler loading compared to N-CC reinforced epoxy and neat epoxy.
- Mechanical properties such as hardness and impact strength of N-C and N-CC reinforced epoxy Nanocomposites increased with increase in filler loading.
- Epoxy reinforced with N-CC Nanocomposites showed superior impact strength among all the composites tested.
- Finally, the results of the investigation confirmed that the mechanical properties especially hardness, tensile and flexural are better for N-C reinforced epoxy Nanocomposites compared to N-CC reinforced epoxy Nanocomposites. Incorporation of N-CC into epoxy helped in improving the impact strength of the composites.

References

[1]Hargitai, H. and Rácz, I., 2012. Applications of Macro-and Microfiller-Reinforced Polymer Composites. Polymer Composites, pp.747-790.

[2]Crawford, R.J., 1998. Plastics engineering. Elsevier.

[3]Hemanth, R., Sekar, M. and Suresha, B., 2014. Effects of fibers and fillers on mechanical properties of thermoplastic composites. Indian Journal of Advances in Chemical Science, 2, pp.28-35.

[4]Veli Deniz, NurselKarakaya and Osman G.Ersoy, (2009) Effects of fillers on the properties of thermoplastic elastomers. Society of Plastics Engineers; 01 – 04.

[5] Katz, H.S. and Mileski, J.V. eds., 1987. Handbook of fillers for plastics. Springer Science & Business Media.

[6]Njuguna, J. and Pielichowski, K., 2003. Polymer Nanocomposites for aerospace applications: properties. Advanced Engineering Materials, 5(11), pp.769-778.

[7]Shenhar, R., Norsten, T.B. and Rotello, V.M., 2005. Polymer-mediated Nanoparticle assembly: structural control and applications. Advanced Materials, 17(6), pp.657-669.

[8]Jung, Y.J., Kar, S., Talapatra, S., Soldano, C., Viswanathan, G., Li, X., Yao, Z., Ou, F.S., Avadhanula, A., Vajtai, R. and Curran, S., 2006. Aligned carbon Nanotube– polymer hybrid architectures for diverse flexible electronic applications. Nano letters, 6(3), pp.413-418.

[9]Fu, S.Y., Feng, X.Q., Lauke, B. and Mai, Y.W., 2008. Effects of particle size, particle/matrix interface adhesion and particle loading on mechanical properties of particulate–polymer composites. Composites Part B: Engineering, 39(6), pp.933-961.

[10]Selvin, T.P., Kuruvilla, J. and Sabu, T., 2004. Mechanical properties of titanium dioxide-filled polystyrene microcomposites. Materials Letters, 58(3-4), pp.281-289

[11]Das, T.K. and Prusty, S., 2013. Graphene-based polymer composites and their applications. Polymer-Plastics Technology and Engineering, 52(4), pp.319-331.

[12]Hammel, E., Tang, X., Trampert, M., Schmitt, T., Mauthner, K., Eder, A. and Pötschke, P., 2004. Carbon Nanofibers for composite applications. Carbon, 42(5-6), pp.1153-1158.

[13]Srivastava, I., 2010. Mechanical properties of polymer Nano-composites. Rensselaer Polytechnic Institute.

[14]Gojny, F.H., Wichmann, M.H., Fiedler, B. and Schulte, K., 2005. Influence of different carbon Nanotubes on the mechanical properties of epoxy matrix composites–a comparative study. Composites Science and Technology, 65(15-16), pp.2300-2313.

[15]Suresha, B., Devarajaiah, R.M., Pasang, T. and Ranganathaiah, C., 2013. Investigation of organo-modified montmorillonite loading effect on the abrasion resistance of hybrid composites. Materials & Design, 47, pp.750-758.

[16]Ray, D., Sengupta, S., Sengupta, S.P., Mohanty, A.K. and Misra, M., 2006. Preparation and properties of vinylester resin/clay Nanocomposites. Macromolecular Materials and Engineering, 291(12), pp.1513-1520.

[17]Azeez, A.A., Rhee, K.Y., Park, S.J. and Hui, D., 2013. Epoxy clay Nanocomposites–processing, properties and applications: A review. Composites Part B: Engineering, 45(1), pp.308-320.

[18]Zhang, M.Q., Rong, M.Z., Zhang, H.B. and Friedrich, K., 2003. Mechanical properties of low Nano-silica filled high density polyethylene composites. Polymer Engineering & Science, 43(2), pp.490-500.

[19]Wetzel, B., Haupert, F. and Zhang, M.Q., 2003. Epoxy Nanocomposites with high mechanical and tribological performance. Composites Science and Technology, 63(14), pp.2055-2067.

[20]Su, F.H., Zhang, Z.Z. and Liu, W.M., 2006. Mechanical and tribological properties of carbon fabric composites filled with several Nano-particulates. Wear, 260(7-8), pp.861-868.

[21]Devaraju, A and Sivasamy,P. "Comparative Analysis of Mechanical Characteristics of Sisal Fibre Composite with and without Nano Particles", *Materials Today: Proceedings, Vol. 5 (*2018), pp.14362-14366.

[22] Devaraju, A, Babu, K and Gnanavel babu, A . "Investigation on the Mechanical properties of Coconut Bunch fiber Reinforced Epoxy with Al_2O_3 Nano particles Composites for Structural Application", *Materials Today: Proceedings, Vol. 5 (*2018), pp.14252-14257.

[23] Devaraju, A., and Pazhanivel, K. "Evaluation of Microstructure, Mechanical and Wear Properties of Aluminium Reinforced with Boron Carbide Nano Composite*", Indian Journal of Science and Technology*, Vol. 9(2), pp. 1-6, 2016.

[24]Rajendran Baskaran, Muthusamy Sarojadevi and ChinnaswamyThangavel Vijayakumar, forced Plastics and Composites 2011 30: 1549 originally published online 5 October 2011.

Experimental Analysis of Heat Transfer Characteristics of Heat Exchanger Using Nano Fluids

T.Mohankumar[A], Dr.K.Rajan[B], K.Sivakumar[C],V.Gopal[D]
[A,C] Research Scholar, Department of Mechanical Engineering,
Bharath Institute of Higher Education and Research, Chennai.
[B]professor, Department of Mechanical Engineering,
Dr.M.G.R.Educational & Research Institution, Maduravayil, Chennai, India.
[D]Associate professor, Department of Mechanical Engineering,
Pallavan college of Engineering, Kanchipuram, India.
Corresponding author: sivakssv@gmail.com

Abstract: The experimental investigation of friction factor and heat transfer characteristics of double pipe heat exchanger using aluminum oxide water based nanofluid. In this present study the heat transfer from hot water to cold water of the double pipe heat exchanger for a plain tube was experimentally investigated. This investigation also to find the heat transfer coefficient, Nusselt number, and Reynolds number. The experimental value of plain tube with water and Al_2O_3 was obtained from the experimental work and these two statistical data were compared with the results. The results shows that the heat transfer rate enhancement of induced with the nanofluids water based flow has give more heat transfer compared to the plain water. The Reynolds numbers were varied in the range of 5000 to 18000. The data obtained from the experimental value were compared with CFD simulation analysis. The performance of heat transfer characteristics of aluminum oxide nanofluid was enhanced compared with smooth plain water used. The experimental value of the heat transfer and friction factor characteristics was well bonding with simulation value of CFD analysis.

Keywords: Heat transfer, Nanofluid, computational fluid dynamics

1. Introduction

Heat exchangers have used for several industrial and engineering applications to improve the performance of heat exchanger devices for enhancement of heat transfer. A majority of heat exchangers used in chemical processing plants, air conditioning equipment, thermal power plants, etc. The heat transfer augmentation techniques are generally classified into three categories namely, Active techniques, and Passive techniques and compared techniques. Sami et al. [1] have investigated Quadrant – cut twisted tape inserts by using CFD analysis of the heat transfer and friction factor characteristic. Al Amin et al.[2] investigating the heat transfer improvement using a rotating twisted tape insert. Date and Saha [3] investigated in a tube fitted with regularly spaced twisted tape elements of the heat transfer. Salam et al. [4] discussed the heat transfer increase using twisted tape insert in a tube. Bodius salam et al [5] have studied the heat transfer enhancement using rectangular- cut twisted tape insert in a tube. Pratik and Warkhedkar [6] were studied about the Heat Transfer Enhancement using Twisted Tape Inserts in a Tube. Durga Prasad et al.[7] investigated the trapezoidal-cut twisted tape insert using Al2O3/water nanofluid in a double pipe u-tube heat exchanger. Sami et al. [8] have discussed with V – cut twisted tape inserts of the numerical investigation of heat transfer. Naga Sarada et al [9] have analyzed the enhancement of heat transfer using twisted tape inserts varying width, were correlation are developed for friction factors and nusselt number for a fully developed turbulent swirl flow. Chang et al [10] have predicting the heat transfer in tube with broken twisted tape inserts with varying twist ratio.

2.Experimental Investigation

The configuration of geometry of the tube with a thickness (t) 0.075cm, length (L) 220cm is used for experimental investigation. In a double pipe heat exchanger as the heat transfer test section to minimize heat loss to the surrounding this is insulated. It consists of two concentric tubes in which hot water flows through the inner tube and cold water flows outer tube in flow through annulus. The outer tube is made of a cast iron having inside and outside diameters of 28mm and 32mm respectively. The inner tube made of an aluminum having inside and outside diameters of 20mm and 18mm respectively. Water is supplied to the pipe from the water loop, an electrical heater controlled by adjusting the voltage, a stirrer and flow meter is placed in the path of the water supply. The temperature sensors were used to measure the outlet temperature of hot and cold fluid and Four thermocouples were used T_1,T_2,T_3,T_4, to find the surface temperature of heat exchanger and data was recorded using data acquisition unit. The cold water is supplied from storage tank into the outer tube and hot water is supplied in the inner tube. The flow meter and pressure gauge were used to measure the mass flow rate of hot fluid and cold fluid flow. The volumetric flow rates of the hot and cold fluid were adjusted by control valves, situated before the inlet ports and experimental reading was taken with different flow rates for obtained the different experimental data. The experimental data were occurred for both plain twisted tape and triangular-cut twisted tape (TCT). The temperature of the inlet and outlet of hot and cold water data's were gathered. The experimental configuration illustrated in Figure 1. The experimental set-up consists of

1. Electrical heater
2. Pressure gauge
3. Temperature indicator
4. Flow meter

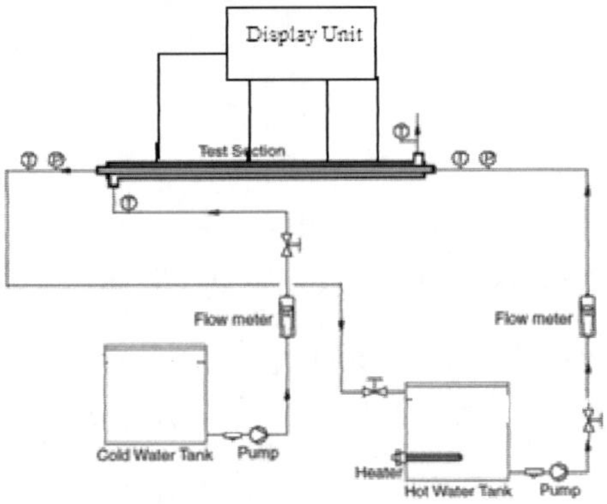

Fig. 1. Experimental setup

Heat transfer rate was calculated

$$Q = m\, Cp\, (T_{out} - T_{in}) \quad \text{---- (1)}$$

Heat transfer coefficient can be calculated

$$h = q / (T_w - T_b) \quad \text{---- (2)}$$

Reynolds number was calculated

$$Re = UD/\nu \qquad \text{---- (3)}$$

Nusselt number can be obtained

$$Nu = h\,D / K \qquad \text{---- (4)}$$

3. Numerical Investigation:

The commercial CFD package was used to perform 3D numerical calculation of the twisted tape of triangular cut and plain twisted tape using the governing equation. The solid flow simulation engineering fluid dynamics is a new breed of computational fluid dynamics software that enables mechanical engineers to simulate fluid flow and heat transfer application. The solid work modeling and meshing was taken and import in to the fluid flow analysis. Numerical values of the mass flow arte and heat transfer used for simulation. In this study, aluminum pipe material were used. The boundary condition for the inlet and outlet of fluid for mesh volume were incorporated. The behavior of the fluid flow in the pipe plain water was studied and the approach by the fluid flow was completely different. Figure 4 shows the numerical simulation for temperature distribution of different flow rates. Figure 5 velocity profiles.

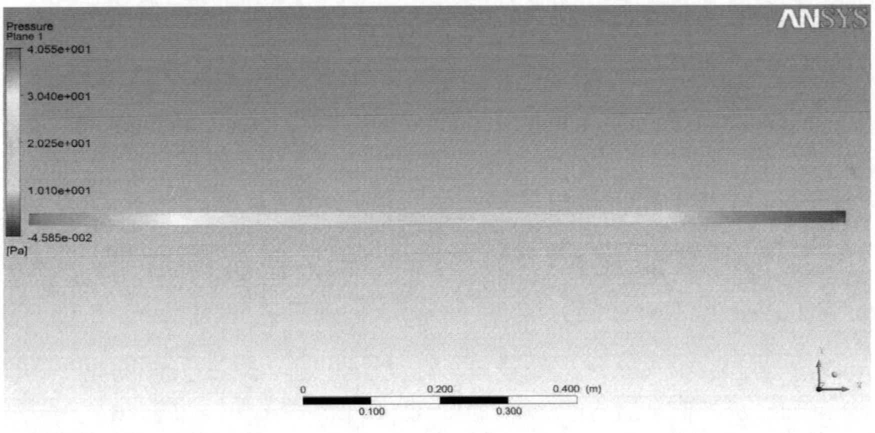

Figure 2 pressure contour

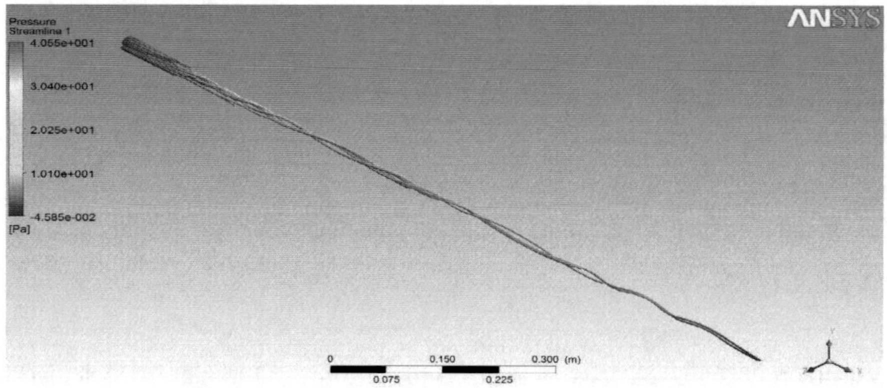

Figure 3 pressure stream line

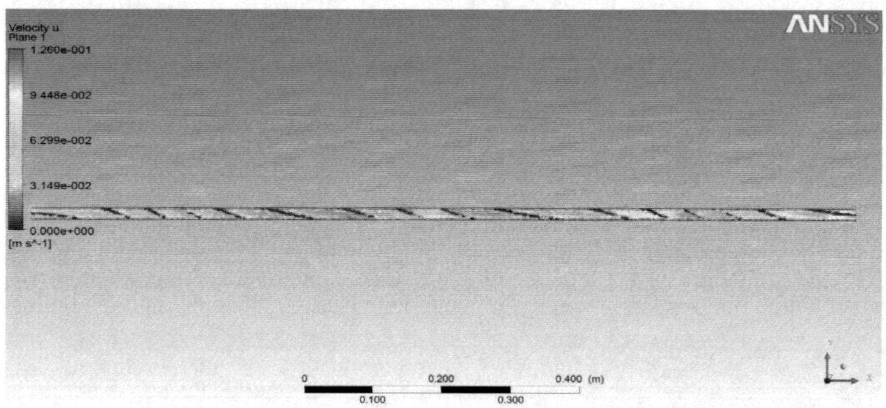

Figure 4 velocity streamline

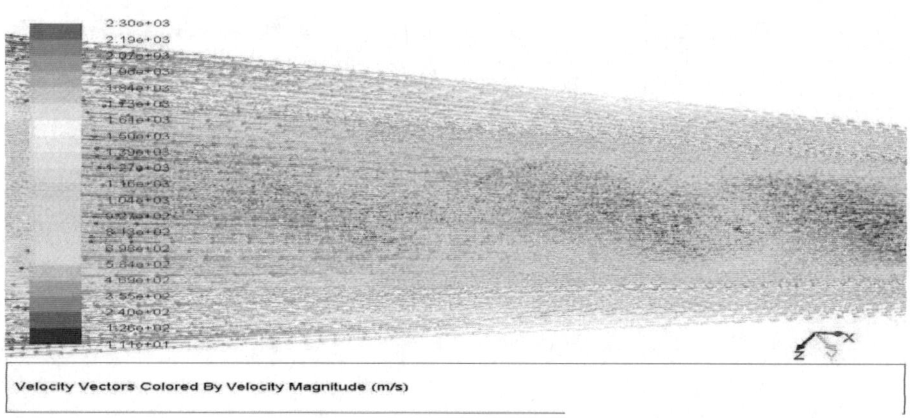

Figure 5 velocity vectors

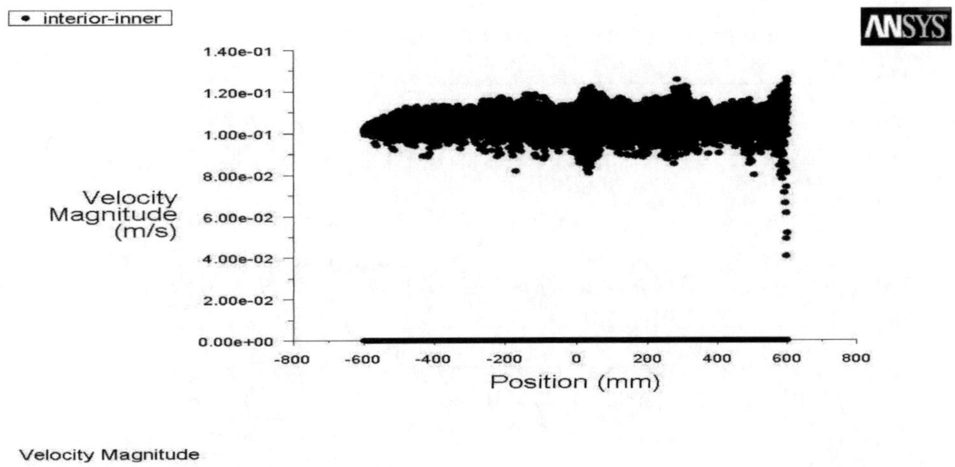

Figure 6 velocity magnitude

4.Result and Discussion

Heat transfer rate characteristics of experimental value for the plain water and nanofluid were collected and these records were taken to check the validity of the numerical simulation value. The Reynolds varied over the range of 5000 to 18000. The graphical representation of numerical value as shown in graph below. The effect on the heat transfer, heat transfer coefficient, Nusselt number, and friction factor were studied numerically. The result of experimental data on the tube fitted with plain water and aluminum oxide are compared with numerical data which is taken from CFD analysis under similar condition. Figure 7 shows the comparison of Reynolds number and Nusselt number of experimental data of plain water and aluminum oxide. It is seen that, the Nusselt number with nano fluids inserts were gives higher values than the plain water inserts. With increases of Reynolds number the Nusselt number also increased. From the observation, the aluminium oxide gives more Nusselt number compared with the plain water used. At the comparable of Re and Nu with nanofluid were improved from 1.02 to 1.19 times compared with the plain water inserts. Figure 8 shows the variation of Reynolds number and heat transfer coefficient, at analogous of nanofluid heat transfer coefficient increased with increases of Reynolds number and gives higher values than the plain water insert. The experimental data of the friction factor and heat transfer of plain water and nanofluids, and these variation with a Reynolds number of nanofluids shown in Figure 9 and 10. It indicates that the heat transfer rate of nanofluid is higher than the plain water insert. For the given Reynolds number, the Nusselt number and heat transfer rate were enhanced with aluminum oxide insert at comparable of plain water inserts.

IOP Conf. Series: Materials Science and Engineering **574** (2019) 012011 doi:10.1088/1757-899X/574/1/012011

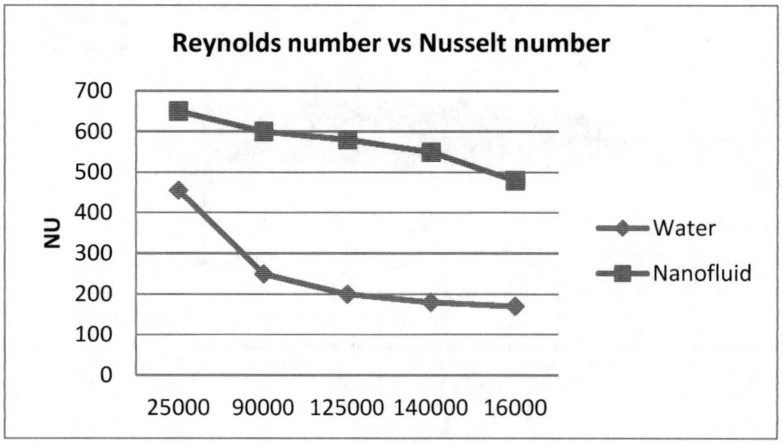

Figure 7.Reynolds number Vs Nusselt number

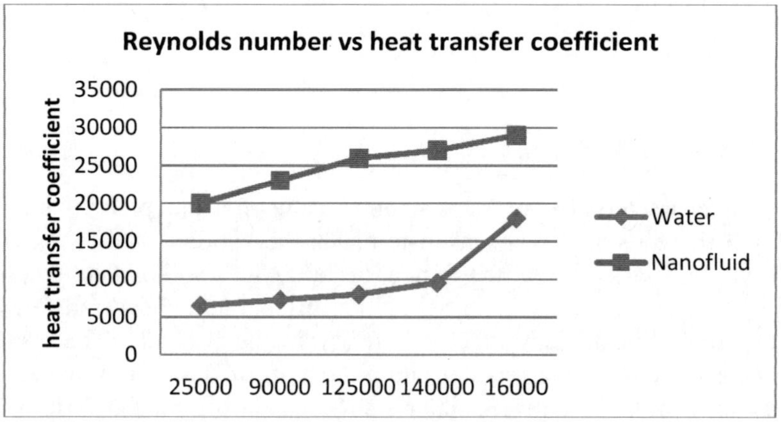

Figure 8.Reynolds number Vs heat transfer coefficient

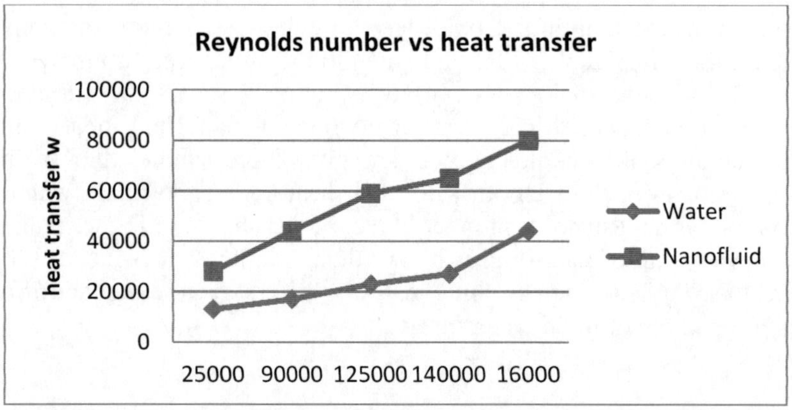

Figure 9. Reynolds number Vs Heat transfer rate

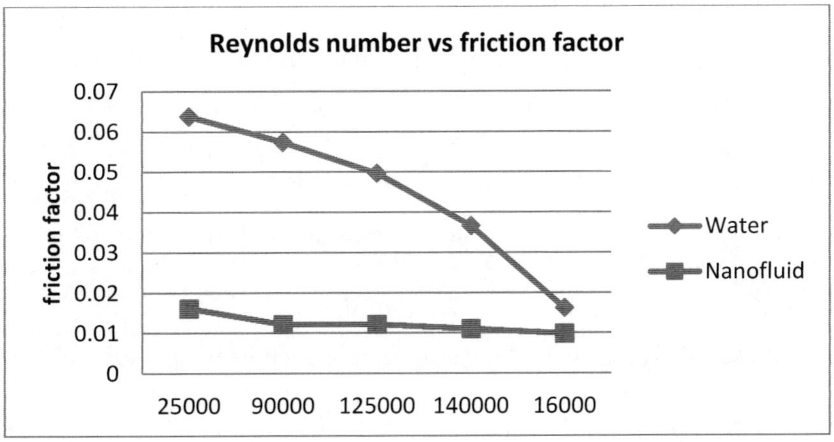

Figure 10. Reynolds number Vs friction factor

5.Conclusion

The heat transfer and friction factor characteristics analyzed have been experimentally performed on the water based Al_2O_3 nanofluid in the pipe. The Results show that the average nusselt number increase with an increase of Reynolds number and the nanoparticle volume concentration. The heat transfer enhancement in the outlet pipes appears larger than the inlet pipes due to the effect of nanofluids. The friction factor reduced with increase of Reynolds number. The Nusselt number was increased with increases of Reynolds number. The experimental values of Nusselt number was increased with 1.02 to 1.19 times. From the experimental results of the plain water and nanofluid, as the heat transfer rate increases by selecting optimum Reynolds number. The present study of nanofluid aluminum oxide insert yields more heat transfer rate, Nusselt number and friction factor were compared with plain water insert. When The nanofluids offered the better performance of heat transfer characteristics as compared with the plain water used. The comparison of numerical value of nanofluid was as well coinciding with the experimental value of nanofluid.

Acknowledgements

The authors would like to thank our Management, principal and Head of the Department and all faculty of the Department Of Mechanical Engineering, for supporting to doing the experimental work in the Thermal Engineering laboratory.

References

1. Sami D Salman, Abdul Amir H Kadhum, Mohd S Takriff, and Abu Bakar Mohamad (2013), CFD analysis of heat transfer and friction factor characteristics in a circular tube fitted with Quadrant Cut twisted tape inserts. Mathematical problems in Engineering, Hindawi publishing corporation, Volume 2013.

2. Al Amin, Zunayad Mohamud, Md Nafis Bin Islam, Md lutfor Rahman, and Dr. Mahammed Ali. Proceedings of 4th Global Engineering, Science and Technology Conference 27-28 December, BIAM Foundation, Dhaka, Bangladesh ISBN: 978-1-922069-43-6.

3. Date, A W and Saha, S K (1993) Numerical Prediction of laminar flow and heat transfer in a tube fitted with regularly spaced twisted tape elements. International journal fluid flow,11: 346-354.

4. Salam B, Biswas S and Bhuiya MMK,(2010), Heat transfer enhancement in a tube using twisted tape inserts. Proc. The 13 th Asian Congress of Fluid Mechanics, 835-838.

5. Bodius Salam , Sumana Biswas , Shuvra Saha and Muhammad Mostafs K Bhuiya (2013), Heat transfer Enhancement in a tube using rectangular – cut twisted tape insert. International conference on Thermal Engineering 5th BSME science direct ,56: 96-103.

6. Pratik P Ganorkar and R M Warkhedkar (2015), Heat Transfer Enhancement in a Tube Using Elliptical-Cut Twisted Tape Inserts. SSRG International Journal of Mechanical Engineering (SSRG-IJME); (2)5.

7. P V Durga Prasad , A V S S K S Gupta and K Deepak (2015), Investigation of Trapezoidal-Cut Twisted Tape Insert in a Double Pipe U-Tube Heat Exchanger using Al2O3/Water nanofluid. 2nd International Conference on Nanomaterials and Technologies (CNT) Science Direct (10), 50 – 63.

8.Sami D Salman, Abdul Amir H Kadhum, Mohd S Takriff, and Abu Bakar Mohamad. Numerical investigation of heat transfer and friction factor characteristics in a circular tube fitted with V- Cut twisted tape inserts. Mathametical problems in Engineering, Hindawi publishing corporation (1).

9. S Naga Sarada, A V Sita Rama Raju , K Kalyani Radha and L Shyam Sunder (2015), Enhancement of Heat transfer using varying width twisted tape inserts. International Journal of Engineering Science and Technology 2015; 6:107-118.

10. Chang S W, Yang T L and Liou J S (2007), Heat transfer and pressure drop in tube with broken twisted tape insert. Experimental Thermal and Fluid Science (32) 2: 489.

IOP Conf. Series: Materials Science and Engineering **574** (2019) 012012 doi:10.1088/1757-899X/574/1/012012

Effect of Wood-Particles Aggregate Percentage on Some Physical Properties of Concrete

Bayan Salim Al Numan[a] , Junaid Kameran Ahmed[b]
[a]*Professor, Tishk International University, Erbil, Iraq.*
[b]*Assistant Lecturer, Tishk International University, Erbil, Iraq.*
Corresponding author: bayan.salim@ishik.edu.iq

Abstract. This paper is focused on light-weight aggregate concrete with wood particles and its issuesaccording to physical and thermal properties. The experimental work for the estimation of density, absorption and thermal conductivity of light-weight wood-particles aggregate concrete is presented. Wide range of wood-particles percentages were used in the experimental work. The thermal conductivity is related to the air-dry density of wood-particles aggregate concrete. The relation is liedclosely within the predictions presented in ACI 213 committee report. The range of air dry density was about 1400 to 1900 kg/m^3.

1. Introduction

There are several advantages of Light-weight aggregate concrete (LWAC) in comparison to normal weight concrete (NWC). The most important is lower bulk density and corresponding lower dead load of a construction that can result in smaller cross-section of construction elements, and buildings foundations. This is especially beneficial for high-rise buildings when lesser loading for columns is needed. The reduction in transportation costs for precast elements is also an advantage.

Applying light-weight concrete in structures reduces construction costs, makes construction easier and leads to green, sustainable construction (Tommy Y. LO et al. 2006).

The light-weight concrete has lower thermal conductivity as well. This ability makes LWAC more suitable for heat insulating constructions and for precast panels. There are also disadvantages. LWAC has higher porosity. It can result in higher absorbing capacity of water and lower compressive strength (Bayan et al. 2004). Light-weight aggregate outer shell thickness, macro porosity and broken grains percentage all affect the aggregate strength (Y. Ke et al. 2009).

A Recent work (I. Asadi et al. 2018) reviewed research works conducted to measure thermal conductivity of normal and light weight concretes. They concluded that using light-weight concrete as building envelopes reduces the amount of heat transfer and energy consumption owing to the lower thermal conductivity of light weight concrete compared to normal weight concrete.

2. Research significance

This paper presents an experimental investigation on thermal conductivity of a sustainable light-weight concrete: wood-particles aggregate concrete. Wide range of wood particles percentages as concrete aggregates were used to investigate the thermal properties which would be beneficial in production of heat insulation components made of wood particle aggregate concrete (Bayan et al. 2004). Some other physical properties of wood-particle aggregate concrete such as density and water absorptions are studied. Wood particles are

IOP Conf. Series: Materials Science and Engineering **574** (2019) 012012 doi:10.1088/1757-899X/574/1/012012

abundant. They are available wherever carpenters are available. Instead of treating them as waste materials, they can be useful in producing sustainable thermal insulated concrete components, especially in precast concrete factories.

3. Experimental work

3.1 Materials: cement, sand, wood particles

Ordinary Portland cement was used in the experiment. The physical and chemical properties of the cement are shown in the tables 1 and 2, respectively. They comply to Iraqi standards [IQS1984].

Table 1. Physical properties of cement

Physical Requirements		Limitation (IQS 1984)	Test Result
Finesse (Blaine)	m^2/kg	Not specified	398
-Initial setting time	min	≥45	180
-Final setting time	hour	≤10	3:45
Soundness (expansion)	mm	≤10	0.0
Compressive strength is not less than (MN/m^2)	2 days	≥20.0	23.0
	28 days	≥42.5	49.9

Table 2. Chemical properties of cement

Chemical Requirements	Limitation (IQS 1984)	Test Result
Sulfate content (as SO3) %	2.5 if C3A ≤5 2.8 if C3A ≥5	2.16
Chloride content %	≤0.10	0.02
Mgnesium oxide (as Mgo) %	≤5.0	3.18
Loss on ignition (Lol) %	Not specified	8.19

River sand was used in the experiment. Table 3 shows the sand properties, and Fig.1 shows the grading (sieve analysis) of the sand. Sand (Table 3, Fig 1) was added in amount such that same consistency shall be gained in every mix. W/C ratio was accordingly varied mixes with incorporated sand, to ensure same workability as the control mix. Table 4 shows the various mix of this work.

Table 3. Sand properties

Properties	Fine aggregate
Specific gravity	2.63
Water absorption	1.07%
Density	1755 Kg/m^3

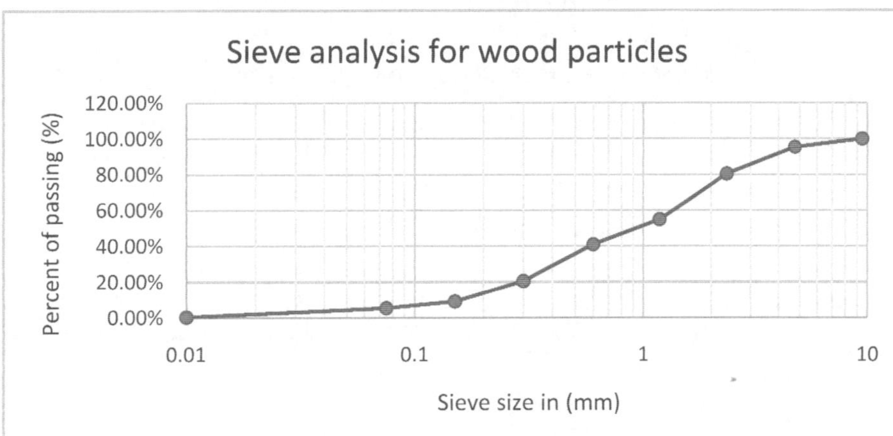

Figure 1.Sieve analysis for wood particles

Wood partical were obtained from the carpentary workshop of Ishik University. The wood is softwood, and the particales were collected from the sawing machine shown in Fig.2 . The coarser particles were taken and the smallest dimension (thickness) distribution is shown in Fig.3. The mean thickness is (1.1) mm. The specificgravity is 0.5.

Figure.2.The sawing machine that the waste particles were collected from

IOP Conf. Series: Materials Science and Engineering **574** (2019) 012012 doi:10.1088/1757-899X/574/1/012012

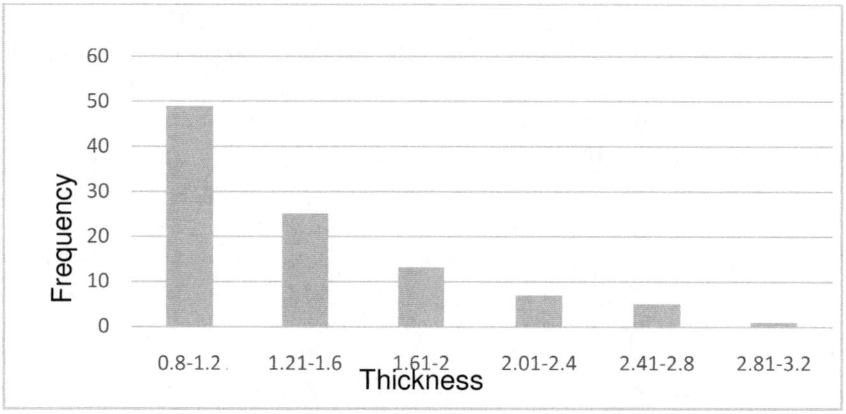

Figure 3. Wood particles size (thickness) distribution

Table 4.Light-weightmix design (mix proportions by mass kg)

	Mix 1	Mix 2	Mix 3	Mix 4	Mix 5
	00	10	20	30	40
Cement	100	90	80	70	60
Sand	200	180	160	140	120
water	50	45	40	35	30
Wood particles	0	9	16	21	24
w/c	0.5	0.5	0.5	0.5	0.5

Light-weight wood-particles concrete mixing method consisted of weighing of materials and mix them dry with hand mixer. Then, adding about half of the required water and mixing continued for 1 minute. Then mixingcontinued with adding the remaining water for 1 minute.

3.2 Specimens

Mixture was cast in prismatic cylindrical molds (40mm diameter and 10mm thickness), for thermal conductivity plates with dimension (40mm diameter and 10mm thickness), were prepared. Molds were filled and then evened with trowel.

Fresh light-weight wood-particle concrete properties were determined. Plunger penetration test method was performed for all five concrete mixes and was tested according to BSEN 1015-4.

3.3 Density of wood particles concrete

Table 5 presents the physical properties of wood particles concrete. Air-dry, dry, and saturated densities are listed for all the tested mixes. Respectively, these were ranged (2340 – 1520), (2112 – 1270), and (2400 – 1680) kg/m^3. Absorption values ranged (13.6% – 32.3%) for control mix to 40% wood particle mix.

3.4 Thermal conductivity

Thermal conductivity is a property of materials that determines the rate of heat flow through the material. Measurement of thermal conductivity involves the flow of heat from one end of the specimen to the other end. The relationship between heat energy per unit time (power) Q and flowing down a specimen of a given cross sectional area A is

$Q = k\,A\,(dt/\,dx)$...(1)

Where, k is the thermal conductivity of the material, W/m. C, and (dt/dx) is the temperature gradient. (Tritt, Terry M)

The test apparatus available in Department of Physics, College of Science, Salahadin University, Erbil, Iraq was used to generate an approximate one-dimensional heat flow through the sample (40 mm in diameter and 10 mm thick) and to achieve the desired mean sample temperatures, in order to obtain the sample thermal conductivity from steady state temperature measurements. The thermal conductivity values for the samples were ranged from 0.609 to 1.900 W/(m C) .

The thermal conductivity values were found displaying an exponential functional relationship with the air-dry density of the woo-particles aggregate concrete, as follows:

$$k = 0.068e^{0.0014w} \quad (2)$$

Figure 4 presents a plot of the results of thermal conductivity against air-dry density. The thermal conductivity ranged from about 0.609 W/(m C) to 1.900 W/(m C). The heat transfer is essentially one-dimensional through the samples.

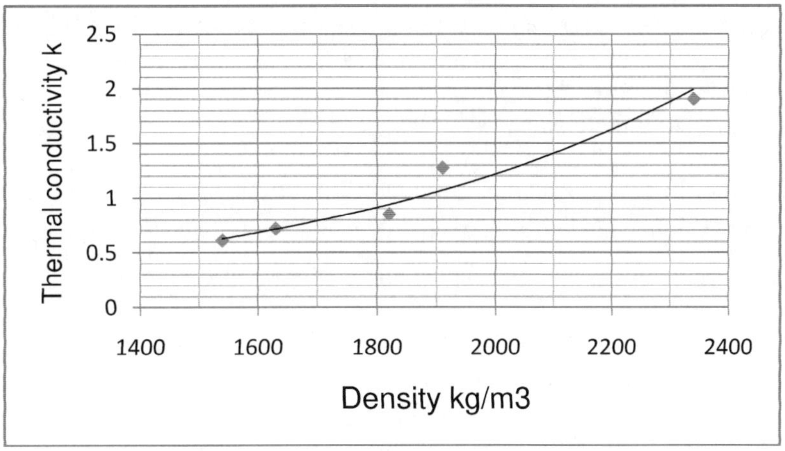

Figure 4.Thermal conductivity- air dry density relationship of wood particles concrete

Table 5 shows the results of air-dry densities and the corresponding thermal conductivity values. Also shown in the table the dry and saturated densities. The water absorption ranged between (%) for control mix to (%) for 40 % wood particle concrete.

Table 5. Physical properties of light-weight concrete with wood particles

Mix	Density (air dry) kg/m^3	Density (dry) kg/m^3	Density (saturated) kg/m^3	Water absorption %	Thermal conductivity W/m C
0	2340	2112	2400	13.6	1.900
10	1910	1800	2084	15.8	1.274
20	1820	1624	1936	19.2	0.849
30	1630	1408	1766	25.4	0.717
40	1520	1270	1680	32.3	0.609

4. Discussion

Light-weight concrete made from wood particles provides mechanical and physical properties which are influenced by selected concrete mix design. The varying of quantity of wood particles in concrete mixresults in a corresponding varying in density, water absorption, and thermal properties of light-weight concrete. Significant difference of light-weight concrete properties between mixes 0 and 40 was observed. Concrete density decreased from 2340 to 1520 kg/m^3, thermal conductivity – from 1.900 to 0.609W/(m·C). Further increase of wood granules amount improved concrete thermal properties considerably. It should be under consideration, with an eye on economy, to use increased quantity of wood particles in mix design for gaining improved concrete thermal properties. water absorption has increased from 13.6% to 32.3% as the density decreased from 2340 to 1520 kg/m^3.

Thermal conductivity has been determined as mentioned in ACI213, for concrete with ovendry density between (320 to 3200 kg/m^3) [ACI 213]. Thermal conductivity values are found related to density of concrete. Fig. 5 shows such relationships for different references along with this work.The range of densities are 800 to 2400 kg/m^3, however, the wood-particles concrete of this work ranged from about 1500 to 1900 kg/m^3 (air-dry), or 1200 to 1800 kg/m^3 (oven-dry). Different relationships are proposed by different investigators. Aggregate type, aggregate microstructure, as well as aggregate grading may be the reasons for these differences. Moreover, method of testing and size of specimen are also affecting the results.

Over 400 test results of density *w* against the logarithm of conductivity k were presented by Valore (1980) and suggested the following equation [ACI 213, 2014]

$$k = 0.072\, e^{0.00125w} \quad (3)$$

For ordinary construction, however, a good estimate of k for concrete may be obtained in the oven-dry condition and, then, it may easily be obtained for air-dry conditions. Thermal conductivity of concrete is increased by the increase in moisture content. Valore (1980) suggested that for each 1% increment in moisture by weight,*k* increases by 6% in relation to ovendry density. The adjusted conductivity is calculated as follows

$k(\text{adjusted}) = k(\text{ovendry}) \times [1 + 6(w_m - w_o) / w_o]$ (4)

where$w_m$andw_o are densities in moist and oven-dry conditions, respectively.
A simple constant factor can be used under conditions of normal protected exposure. Thethermal conductivityvalues, when adjusted for moisture in ordinary exposure, are to be increased by 20% over standard values for oven-dry concrete. This results in adjustingValore equation (equation 1), which now becomes(ACI 213, 2014):

$k = 0.086\ e^{0.00125w}$ (5)

Asadi et al (2018) proposed a relationship of thermal conductivity-concrete density based on 185 experimental date available in literature for density range of (150 – 2350 kg/m^3), as follows:

$k = 0.0625\ e^{0.0015w}$ ($R^2 = 0.81$) (6)

The proposed equation of this work related to wood particles light-weight concrete (equation 2) is shown in Fig. (5)together with ACI 213 equation, and Asadi equation. A comparison between these equations and the experimental equation proposed in this study clearly shows close results for the range of air-dry densities used in this work (1520 – 2340) kg/m^3.

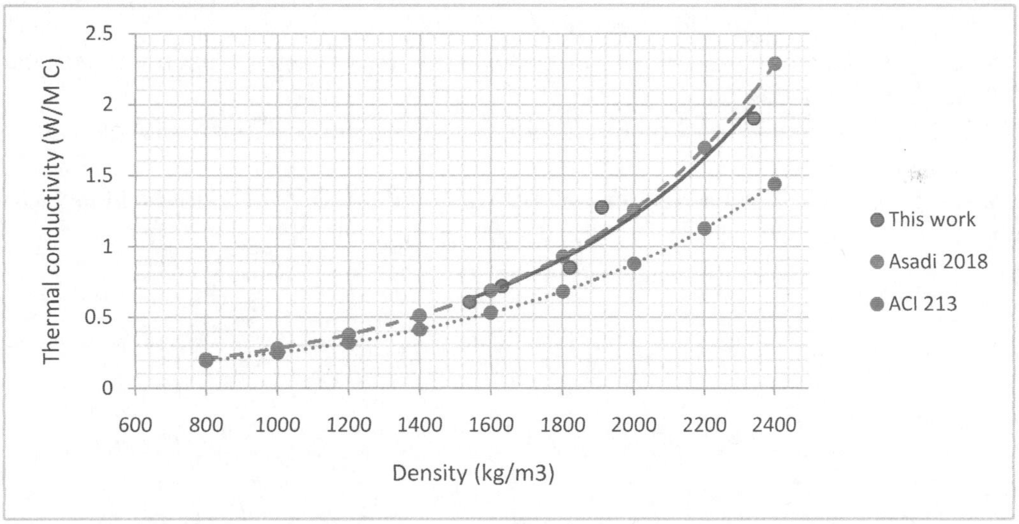

Figure 5. Equation developed for wood particles concrete relating thermal conductivity and density, compared with ACI 213 R-14 and Asadi 2108

Conclusions

The light-weight concrete properties are influenced by the amount of wood granules in mix design. By varying the amount of wood particles in light-weight concrete mixvariouslight-weight concrete properties can be resulted.

IOP Conf. Series: Materials Science and Engineering **574** (2019) 012012 doi:10.1088/1757-899X/574/1/012012

Thermal conductivity increases with the increasing of wood particles amount. Light-weight concrete made with wood particles can be used as an insulationmaterial with low thermal conductivity. The thermal conductivity of light-weight concrete with wood particles is depended upon the thermal properties of cement paste and the wood particle properties.

It can be concluded that further increase of wood particle amount in mix design provides reduction in concrete thermal conductivity which is important for sustainability reasons.

By increasing wood particles amount in light-weight concrete its density could be reduced from 2340 to 1520 kg/m^3 and thermal conductivity from 1.900 to 0.609W/(m·C).

References

[1] Bayan S. Al Numan, Ayad A. Slaby and AbdulQader I. Al Hadithi, "Some mechanical properties of sawdust concrete", Iraqi Journal of Civil Engineering, March 2004, pp27 – 34.
[2] Tritt, Terry M. Thermal Conductivity: Theory, Properties, and Applications. Kluwer Academic/ Plenum Publishers. New York, 2004.
[3] Ke, Y., Beaucor, A. L., Ortola, S., Dumontet, H., Cabrillac, R. 2009. Influence of volume fraction and characteristics of light-weight aggregates on the mechanical properties of concrete. In: Construction and Building Materials 23. pp. 2821– 2828. http://dx.doi.org/10.1016/j.conbuildmat.2009.02.038
[4] Tommy Y. Lo, Tang, W. C., Cui, H. Z. The effects of aggregate properties on light-weight concrete. Building and Environment 42. 2007.
[5] Valore (1980). Calculating U-values of hollow concrete masonry. Concrete International, V.2, No. 2, Feb. pp 40 – 63.
[6] ACI 213R (2014) Guide for StructuralLight-weight-Aggregate Concrete. American Concrete Institute. 49p.
[7] I. Asadi, P. Shafigh, Z. Abu Hassan, and N. Mahyuddin (2018). Thermal conductivity of concrete – a review. Journal of Building Engineering 20, pp.81 – 93.

Characterization of New Natural Cellulosic Fibers – A Comprehensive Review

Arul Marcel Moshi.A*, Ravindran.D, Sundara Bharathi.S.R, Suganthan.V and Kennady Shaju Singh.G
Department of Mechanical Engineering, National Engineering College, Kovilpatti – 628 503, TamilNadu, India.
*Corresponding author – moshibeo2010@gmail.com

Abstract. The world is in the need of more eco-friendly materials; therefore researchers around the globe focus on developing new materials that would improve the environmental quality of products. This need for new green materials has led to the utilization of composites made from raw natural fibers as they are low cost and have better mechanical properties and require low production energy consumption. Natural fibers are one of the effective substitutes for switching artificial fiber and concentrating to reinforce polymer matrices due to their decomposable characteristic. This study explains the various tests to be carried out such as mechanical property analysis, chemical composition analysis, surface morphology analysis, thermal stability analysis, etc., in order to characterize a new natural fiber. The impact of treating natural fibers with the suitable chemicals on the desirable properties is elaborately discussed. The testing procedures and the various significant points given by the researchers in this field are presented in detail. This work will surely be helpful for the upcoming researchers those who would like to carry out the research work in natural fiber composites.

1. Introduction

Researchers are doing variety of investigations on different types of application areas of composite structures for various applications. For performing the desirable analysis on the composites structures to ensure their suitability, it is very essential to have the details of the base properties of the constituents of the composites. Hence, characterization study holds most significant role while we are introducing new natural fibers to this field. Standard testing procedures are being followed for carrying out the property analysis of natural fibers. The consolidated details obtained from the extensive literature study are presented in this work. P. G. Baskaran *et al.*, have analysed the mechanical properties, chemical composition, crystalline characteristics and thermal degradation of Dichrostachys Cinerea Fibre (DCFs). The mechanical properties of the DCFs proved that they are the best replacement for manmade synthetic fibres in polymer matrix composites [1]. R. Gopinath *et al.*, have investigated the chemical composition, crystalline characteristics, thermal degradation, and mechanical properties of Sida Rhombifolia Fibre (SRFs). The test results revealed the presence of high cellulose content (75.09 wt. %,) in SRFs, considerably more significant than that of other cellulosic fibres [2]. N. Rajesh Jesudoss Hynes *et al.*, have described the chemical configuration, crystalline characteristic, and thermal stability of the fiber obtained from Heteropogon contortus Plant (HCFs). The reasonably higher percentage of cellulose (64.875%) in HCFs can offer higher strength and the lesser density (602 kg/m^3) which make HCFs as fit for low weight composite applications [3]. S. Indran *et al.*, reported the characterisation details of Cissus Quadrangularis Stem Fibre (CQSF). The density of CQSF is significantly lower than that of the popular glass fibre, carbon fibre, etc. and also gives 20% better specific strength. The tensile strength and Young's moduli are 4203 ± 1276 MPa and 131 ± 37 GPa which is significantly higher than any other natural fibres ever investigated and also on bar with the widely used synthetic fibres [4]. J. Jayaramudu *et al.*, in their research work stated the properties of the uniaxial natural fabric G. tilifolia. The SEM analysis manifested the morphology of the fiber contains uniaxial roughly parallel surfaces present in th fiber [5].

MabroukMaache *et al.*, have investigated the properties of a new lignocellulosic fibre extracted from Juncus Effusus L (JE). The study of the surface morphology by SEM revealed that the cross-section of the JE fibre bundle has a cellular shape similar to other fibres [6]. C.E. Maepa *et al.*, have described the extraction and characterisation of Tassel Fibre. The authors stated that the residue from maize (tassel fibres) is an attractive substitute as a cellulose resource for some applications [7]. M. V. Maheshwaran *et al.*, have reported the chemical, structural, thermal, and tensile properties of the Epipremnum Aureum stem (EAFs). It has been shown that lignin content has little effect on biological attack; and relatively lesser density of the EAFs may enable them for light-weight applications [8]. M. Prithiviraj *et al.*, have studied the mechanical properties of the fibre, chemical composition, crystalline characteristics and thermal degradation of Perotis Indica Plant (PIFs). The obtained properties of the PIFs ensured that they are suitable as a good substitute for the conventional materials in the industries [9]. Dhana Lakshmi SampathKumar *et al.*, have reported that the Chemical treatment of natural lingo-cellulosic fibres would remove the impurities like pectin, fat and lignin present in the fibres, which may result in rough surfaces in turn the better ability of bonding of fiber with the matrix while preparing composite structures with the help of them. From the literature survey, we found that the natural fibres will be more important in future because of its availability, recyclability and eco-friendly nature. Various natural cellulosic fibres have been investigated which impart different desirable properties.

2. Characterization study

For the characterization study, initially a new natural fiber needs to be identified from a plant. Then, it should be separated from the plant by following the commonly used step by step procedure as described below [11, 12]:

i. The part of the plant (Ex.: stem, root, etc,) from which the fiber is going to be separated are collected at required level.
ii. The collected parts of the plant are needed to be immersed in ordinary / distilled water for about two to three weeks for getting microbial degradation.
iii. Then the fiber is extracted from the stem / root with the help of a metallic comb.
iv. After that, the fiber is washed in running water in order to remove the unwanted particles present in the fiber surface.
v. Followed by, the fiber is dried in oven at nominal temperature to remove the moisture content.

2.1. Physical Analysis

Physical analysis is normally performed to predict the density and nominal geometrical details of the chosen fiber. Density is normally measured by using gas pycnometer set up and

IOP Conf. Series: Materials Science and Engineering **574** (2019) 012013 doi:10.1088/1757-899X/574/1/012013

diameter is measured by means of micrometer. The procedure to measure the density value is as follows [5]:

- The gas pycnometer operates by detecting the mass difference between the known and unknown density materials.
- Toluene is normally chosen as the known density (867 kg/m^3) liquid for the density measurement of fibers.
- A small amount of the fiber needed to be tested in chopped form and a little amount of toluene are taken for the analysis. The mass of each are measured.
- Distilled water is also sometimes used as the known density liquid [6].
- The mass values are to be noted for the empty pycnometer, chopped fiber only, toluene only and toluene + chopped fiber.
- Then, the density of the fiber can be measured by means of the following formula:

$$\rho_f = \{\frac{m_2 - m_1}{(m_3 - m_1)(m_4 - m_2)}\}\rho_t$$

Where, m_1-Mass of empty pycnometer

m_2- mass of pycnometer with chopped fibre

m_3- mass of pycnometer filled with toluene

m_4- mass of pycnometer filled with chopped fibre and toluene solution

ρ_f – density of fibre

ρ_t – density of toluene

The measured density values of various natural fibers are presented in table 1.

Table 1. Density values of various natural fibers [1, 3, 6, and 7]

S.No.	Name of the fiber	Density (kg/m^3)
1.	Dichrostachys cinerea	1240
2.	Prosopis juliflora	580
3.	Acacia Leucophloea	1385
4.	Nettle	1510
5.	Cyperus pangorei	1102
6.	Lygeum Spartum	1499.7
7.	Sida Rhobifolia stem	1320.7
8.	Jute	1460
9.	Flax	1500
10.	Ramie	1500
11.	Hemp	1480
12.	Kenaf	1400
13.	Sisal	1500
14.	Cotton	1600
15.	Heteropogan Contortus	602
16.	Acacia planiforns	660
17.	Coir	1200
18.	Pineapple leaf	1400
19.	Banana	1350
20.	Palmyrah	1090

21.	Bamboo	910
22.	Bagasee	1200
23.	Sea grass	1500

2.2. Chemical Analysis

Chemical composition analysis is performed to identify the percentage composition of cellulose contents, hemicellulose contents, lignin contents, moisture, wax and ash contents present in the fiber. Since the percentage existence of above said compositions decides the strength and bonding ability of the fiber materials, this study is significant in characterization work [7, 13, 14].

Determination of cellulose content – The amount of cellulose present in the fiber is found out by Kurschner's and Hanack's method. This method is based on the insolubility of cellulose in water and its resistance to the action of dilute acids and bases.The crushed fibre sample was degraded with a mixture of nitric acid and acetic acid and boiled in an apparatus that contained a condenser. The solution is then filtered through a Büchner funnel. Then, the filter paper containing an insoluble residue is dried in an oven and measured which is the direct measure of the cellulose content [6, 15].

Determination of lignin content - Insoluble solution of the lignin content is determined based on Klason lignin of APPITA P11s-78 method. The samples of crushed fibres (300 mg) are hydrolysed with 72% of sulfuric acid in an ultrasonic bath for 1 hour at a controlled temperature of 30 °C. The treated sample is mixed with a solvent solution of methylene chloride (CH2Cl2) and placed in an autoclave at 125 °C for one hour and dried. Then, the sample was cooled; the lignin was filtered and measured [5, 16].

Determination of ash content – The ash measurement test is carried out basd on ASTME1755-61 standard. The amount of ash present in the fiber is predicted based on the difference in weight between the burned and unburned fibers [4, 17].

Determination of moisture content – The amount of moisture content is measured by means of the standard weight loss method. Or that, the test sample is heated until there is no weight difference can be noted. At that condition, the weight difference of the fiber is noted which is the amount of moisture content present in the fiber [3, 18].

Determination of hemicellulose content – This test is carried out based on the NFT standard 12-008 [2, 3, 4].

Determination of wax content – This measurement is based on the Conrad method [3, 4].

For achieving better bonding ability and mechanical strength, the percentage of cellulose content should be high, hemicellulose content should be low, lignin content should be nominal, moisture, ash and wax contents should be low [1, 2, 19].

The chemical compositions of various investigated natural fibers are presented in table 2.

Table 2. Chemical composition of few natural fibers [1, 2, 4, 6, 8]

S.No.	Name of the fiber	% of Cellulose	% of Hemicellulose	% of Lignin	% of moisture	% of Ash	% of Wax
1.	Dichrostachys Cinerea	72.4	13.08	16.89	9.82	3.97	0.57
2.	Prosopis Juliflora	61.65	16.14	17.11	9.48	5.2	0.61
3.	Acacia Leucophloea	68.09	13.6	17.73	8.83	0.08	0.55
4.	Sida Rhombifolia	75.09	15.43	7.48	12.02	4.07	0.49

	Stem						
5.	Acacia leucophloea	68.09	13.6	17.73	8.83	0.08	0.55
6.	Jute	72	13	13	12.6	1	0.5
7.	Flax	81	14	3	10	--	1.7
8.	Hemp	74	18	4	10.8	--	2.3
9.	Kenaf	53.14	14.33	8.18	9	3.5	0.8
10.	Sisal	68	13	11.5	17	--	2
11.	Heteropogon Contortus	64.87	19.34	13.56	7.4	--	0.22
12.	Acacia Planiforns	73.1	9.41	12.04	8.21	4.06	0.57
13.	Acacia Arabica	68.10	9.36	16.86	--	--	0.49
14.	Cissus Quadrangularis Stem	82.73	7.96	11.27	6.6	--	0.18
15.	Cissus Quadrangularis root	77.17	11.02	10.45	7.3	--	0.14
16.	Agave	68.42	4.85	4.85	7.69	--	0.26
17.	Epipremnum Aureum	66.34	13.42	14.01	7.41	4.61	0.37
18.	Perotisindica	68.4	15.7	8.35	9.54	4.32	0.32
19.	Red Banana Peduncle	72.90	11.01	15.99	9.36	2.79	0.32
20.	Furcraea Foetida	68.35	11.46	12.32	5.43	6.53	0.24

2.3. Thermogravimetric Analysis

Thermogravimetric analysis is carried out on fibers in order to find out their thermal stability in turn to identify the suitable high temperature applications. The test results will provide the temperature points at which the various constituents such as cellulose, hemicellulose, lignin, moisture etc., present in the fiber getting degraded. The analysis is performed in Jupiter simultaneous thermal analyzer in which an alumina crucible will be there to keep the sample. About 12 mg amount of sample fiber to be tested is taken in the crucible and it is subjected to a varying temperature condition from 15 to 600^0 Celsius under the atmosphere of Nitrogen gas. The readings are taken for the heating rate of 10^0C/minute [20, 21, 22].

The thermal stability of various constituents of different natural fibers is illustrated in table 3.

Table 3. Thermal stability values of the constituents of few natural fibers [1, 3, 5, 7]

S.No.	Name of the fiber	Moisture (^{0}C)	Hemicellulose (^{0}C)	Cellulose (^{0}C)	Wax (^{0}C)
1.	Dichrostachys Cinerea	164.4	284.3	359.3	495.3
2.	Lygeum Spartum	110	220 – 307	338.7	450 – 550
3.	Sida Rhombifolia stem	73.5	250 – 375	328.2	433.3
4.	Heteropogon Contortus	90.1	240 – 400	337.7	500 – 650
5.	Cissus Quadrangularis stem	89	294	342	476 – 695
6.	Juncus Effusus	30 – 110	220 – 360	300	400 – 500

7.	Epipremnum Aureum	110	240	240 – 380	380 – 600
8.	Perotis Indica	88.4	277.5	339.1	400 – 500
9.	Areca	241.5	241.5 – 297	297 – 353	499.4
10.	Cissus Quadrangularis root	89	230 – 330	328.9	481 – 696
11.	Red Banana Penduncle	120	200 – 300	353	353 – 500
12.	Areca Palm Leaf Stalk	60 – 110	200 – 350	365	37 – 450

2.4. Fourier Transform Infrared (FTIR) Analysis

FTIR analysis is carried out to discover the chemical functional group present in the natural fiber to ensure the existence of the chemical components. This test is conducted in a FTIR spectroscopy. For the test, 2.5 mg quantum of the fiber sample is mixed with potassium bromide with the ratio of 1:100 ratio and formed as pellets. The test samples will be subjected to infrared radiation and the frequency responses are noted down. The resultant plot obtained out of the analysis has been plotted between wave number and normalized transmittance, which carries a frequency wave with critical peaks. The critical peaks ensures the existence of O-H group in turn cellulose contents, C-H group in turn hemicellulose content, C=C group in turn wax content and C-OH group in turn lignin content [6, 23, 24].

2.5. X-Ray Diffraction (XRD) Analysis

XRD analysis is performed to predict the crystallinity index value of the fiber. Crystalline index represents the degree of crystalline nature of the fiber. Crystallinity index is calculated by the following formula [1, 2, 6]:

$$\text{Crystallinity Index, CI} = \left(1 - \frac{\text{Min .peak intensity representing the amorphous region}}{\text{Max .peak intensity representing the crystalline region}}\right) \times 100\%$$

The crystalline size is predicted with the help of Scherer's formula:

$$\text{Crystallite Size, CS} = \frac{0.89 \times \text{Amplitude of the radiation}}{\beta \cos \theta}$$

Where, 0.89 – Scherer's constant,

β – full width of the peak at half – maximum, and

2θ – Diffraction angle.

The crystallinity index values reported for different natural fibers are provided in table 4.

Table 4. Crystallinity Index values for few natural fibers [2, 4, 6, 7]

S.No.	Name of the fiber	Crystallinity Index (%)
1.	Dichrostachys Cinerea	57.82
2.	Grewia Tilifolia	41.7
3.	Prosopis Juliflora	46
4.	Lygeum Spartum	46.19
5.	Date Palm	38.5
6.	Coconut	41.9
7.	Wrightia Tinctoria	40.6
8.	Sansevieria Ehrenbergii	52.27
9.	Sansevieria Cylindrica	60
10.	Altaea Officinalis	68
11.	Sida Rhombifolia	56.6

12.	Ferula	48
13.	Acacia Leucophloea	51
14.	Heteropogan Contortus	54.1
15.	Cyperus Pangorei	41
16.	Acacia Arabica	51.72
17.	Acacia Planiforns	65.38
18.	Flax	70
19.	Ramie	58
20.	Sisal	75
21.	Coconut	68
22.	Hemp	87.9
23.	Jute	58.9
24.	Pineapple	58.6

2.6. Morphology Analysis

Morphology study is done to know the nature of the surface of the fiber, which is performed with Scanning Electron Microscopy. From the SEM results, it can be understood the distribution of various constituents present over the fiber surface. The presence of hemicellulose contents can be viewed as bright white layers. The existence of lignin content can be noted as nail like parts. The wax and impurity items will be noted by the small discontinuous parts. By observing the nature of the surface, i.e, either rough or smooth, one can come to know the conclusion about the bonding nature of the fiber under analysis with the matrix medium. Chemical treatments will help to make the fiber surface as rough in turn to make the fiber with more bonding ability [4, 9, 25].

2.7. EDX Analysis

The amount of elements present on the fiber's surface (such as Carbon, Oxygen, Nitrogen, etc.,) is tested using Energy Dispersive X-Ray Spectroscopy [2, 5, 6].

2.8. Mechanical Strength Test

In order to find out the tensile strength of the natural fibers, INSTRON 5500 type tensile testing machine is normally used [3, 4 7]. The test is usually conducted at room temperature condition. The tensile strength vales reported for various natural fibers are listed in table 5.

Table 5. Tensile strength values of few natural fibers [2, 6, 8, 9]

S.No.	Name of the fiber	Tensile strength (MPa)
1.	Dichrostachys Cinerea	873 ± 14
2.	Lygeum Spartum	280
3.	Sansevieria	584.64 – 724.12
4.	Banana	529 – 759
5.	Jowar	302
6.	Sisal	274 – 526
7.	Alfa	134 – 220
8.	Kenaf Bast	427 – 519
9.	Petiole Bark	185.52
10.	Spatha	75.66

11.	Arundo Donax	248
12.	Grewia Tilifolia	75.3
13.	Agave	63 – 211
14.	Elephant Grass	292.3
15.	Empty Fruit Bunche	49
16.	Date palm	170 – 275
17.	Sida Rhombifolia stem	673 ± 14
18.	Heteropogan Contortus	476 ± 11.6
19.	Jute	400 – 773
20.	Flax	500 – 1500
21.	Cotton	200 – 400
22.	Coir	95 – 174
23.	Ramie	400 – 938
24.	Hemp	690
25.	Pineapple	362 – 748
26.	Kenaf	930
27.	Sisal	511 – 635
28.	Banana	700 – 800

3. Conclusion

The importance of characterization of natural fibers and the treatment of fibers with suitable chemicals for improving the strength are clearly explained in this work.

- Various tests to be conducted in order to characterize a new natural fiber are detailed with the processing procedure.
- It is also explained that how to interpret the results obtained from each test.
- The mass density, chemical composition, thermal stability, crystallinity index values and tensile strength results of distinct natural fibers are consolidated and presented in tables.
- This work may be an initiative for many future researchers those who are going to work with natural fibers and fiber reinforced polymer composite structures.

REFERENCES

[1] P. G. Baskaran, M. Kathiresan, P. Senthamaraikannan& S. S. Saravanakumar, “Characterization of New Natural Cellulosic Fibrefrom the Bark of Dichrostachys Cinerea”, Journal of NaturalFibres 15(1) (2017), pp: 62-68, DOI: 10.1080/15440478.2017.1304314.

[2] R. Gopinath, K. Ganesan, S. S. Saravanakumar & R. Poopathi, “Characterization of new cellulosic fibre from the stem of Sida rhombifolia”, International Journal of Polymer Analysis and Characterization 21(2) (2016), pp: 123-129, DOI: 10.1080/1023666X.2016.1117712.

[3] N. Rajesh Jesudoss Hynes, N. J. Vignesh, P. Senthamaraikannan, S. S. Saravanakumar & M. R. Sanjay, “Characterization of New Natural Cellulosic Fibre from Heteropogon

contortus Plant", Journal of Natural Fibres 15(1) (2017), pp: 146-153, DOI: 10.1080/15440478.2017.1321516.

[4] S. Indran, R. Edwin Raj, "Characterization of new natural cellulosic fibre from Cissusquadrangularis stem", Carbohydrate Polymers 117 (2015), pp: 392–399, DOI: doi.org/10.1016/j.carbpol.2014.09.072.

[5] J. Jayaramudu, B.R. Guduri, A. VaradaRajulu, "Characterization of new natural cellulosic fabric Grewiatilifolia", Carbohydrate Polymers 79 (2010), pp: 847-851, DOI:10.1016/j.carbpol.2009.10.046.

[6] Mabrouk Maache, Abderrezak Bezazi , Salah Amrounea,, Fabrizio Scarpa, Alain Dufresne, "Characterization of a novel natural cellulosic fibre from Juncuseffusus L", Carbohydrate Polymers 171 (2017), pp: 163–172, DOI:https://doi.org/10.1016/j.carbpol.2017.04.096.

[7] C. E. Maepa, J. Jayaramudu, J. O. Okonkwo, S. S. Ray, E. R. Sadiku& J. Ramontja, "Extraction and Characterization of Natural Cellulose Fibres from Maize Tassel", International Journal of Polymer Analysis and Characterization 20(2) (2015), pp: 99-109, DOI:10.1080/1023666X.2014.961118.

[8] M. V. Maheshwaran, N. Rajesh JesudossHynes, P. Senthamaraikannan, S. S. Saravanakumar& M. R. Sanjay, "Characterization of natural cellulosic fibre from Epipremnumaureum stem", Journal of Natural Fibres 15(6) (2017), pp: 789-798, DOI: 10.1080/15440478.2017.1364205.

[9] M. Prithviraj, R. Muralikannan, P. Senthamaraikannan& S. S. Saravanakumar, "Characterization of New Natural Cellulosic Fibre From PerotisIndica Plant", International Journal of Polymer Analysis and Characterization 21(8) (2016), pp: 669-674, DOI: 10.1080/1023666X.2016.1202466.

[10] Dhanalakshmi Sampathkumar, Ramadevi Punyamurthy, Basavaraju Bennehallic, Srinivasa Chikkol Venkateshappa, "Physical characterisation of natural lingo cellulosic single areca fibre", Ciência&Tecnologia dos Materiais 27(2) (2015), pp: 121–135, DOI: https://doi.org/10.1016/j.ctmat.2015.10.001.

[11] Arthanarieswaran, V. P., A. Kumaravel, and S. S. Saravanakumar, "Characterization of new natural cellulosic fiber from acacia leucophloea bark", International Journal of Polymer Analysis and Characterization 20 (2015), pp: 367–76. DOI:10.1080/1023666X.2015.1018737.

[12] Arthanarieswaran, V. P., A. Kumaravel, and S. S. Saravanakumar, "Physico-chemical properties of alkali-treated acacia leucophloea fibers", International Journal of Polymer Analysis and Characterization 20(8) (2015), pp: 704–13. DOI:10.1080/1023666X.2015.1081133.

[13] Ashok, B., K. Obi Reddy, K. Madhukar, J. Cai, L. Zhang, and A. Varada Rajulu, "Properties of cellulose/Thespesia lampas short fibers bio-composite films", Carbohydrate Polymers 127 (2015), pp:110–115. DOI:10.1016/j.carbpol.2015.03.054.

[14] Malkapuram, R., V. Kumar, and S. N. Yuvraj, "Recent development in natural fibre reinforced polypropylene composites', Journal of Reinforced Plastics and Composites 28 (2009), pp: 1169–1189. DOI: 10.1177/0731684407087759.

[15] Mayandi, K., N. Rajini, P. Pitchipoo, J. T. Winowlin Jappes, and A. Varada Rajulu, "Extraction and characterization of new natural lignocellulosic fiber Cyperus pangorei" International Journal of Polymer Analysis and Characterization 21 (2016), pp: 175–83. DOI:10.1080/1023666X.2016.1132064.

[16] Mitchell, C., M. Carr, M. Parfitt, J. C. Vickerman, and C. Jones, "Surface chemical analysis of raw cotton fibers and associated materials. Cellulose 12 (2005), pp: 629–39. DOI: 10.1007/s10570-005-9000-9.

[17] Obi Reddy, C., E. Uma Maheswari, E. Muzenda, M. Shukla, and A. Varada Rajulu, Extraction and Characterization of Cellulose from Pretreated Ficus (Peepal Tree) Leaf Fibers", Journal of Natural Fibers 13 (2016), pp: 54–64. DOI:10.1080/15440478.2014.984055.

[18] Obi Reddy, K., K. Raja Narender Reddy, Z. Zhang, J. Zhang, and A. Varada Rajulu, "Effect of alkali treatment on the properties of century fiber", Journal of Natural Fibers 10 (2013), pp: 282–296. DOI:10.1080/15440478.2013.800812.

[19] Reddy, N., and Y. Yang, "Characterizing natural cellulose fibers from velvet leaf (Abutilon theophrasti) stems", Bioresource Technology 99 (2008), pp: 2449–2454. DOI:10.1016/j.biortech.2007.04.065.

[20] Sathishkumar, T. P., P. Navaneethakrishnan, and S. Shankar, "Tensile and flexural properties of snake grass natural fiber reinforced isophthallic polyester composites", Composites Science and Technology 72 (2012), pp: 1183–1190. DOI:10.1016/j.compscitech.2012.04.001.

[21] Sreenivasan, V. S., S. Somasundaram, D. Ravindran, V. Manikandan, and R. Narayanasamy, "Microstructural, physico-chemical and mechanical characterisation of Sansevieria cylindrica fibres—An exploratory investigation", Materials & Design 32 (2011), pp: 453–461. DOI:10.1016/j.matdes.2010.06.004.

[22] Thakur, V., and A. S. Singha, Natural fibers-based polymers: Part I-Mechanical analysis of Pine needles reinforced biocomposites', Bulletin Materials Sciences 33 (2010), pp: 257–64. DOI: 10.1007/s12034-010-0040-x.

[23] Subramanian, K., Kumar, P. S., Jeyapal, P., & Venkatesh, N., "Characterization of lingo cellulosic seed fibre from Wrightia Tinctoria plant for textile applications—an exploratory investigation", European Polymer Journal, 41(4) (2005), pp: 853-861. DOI: 10.1016/j.eurpolymj.2004.10.037

[24] Lu, T., M. Jiang, Z. Jiang, D. Hui, Z. Wang, and Z. Zhou, "Effect of surface modification of bamboo cellulose fibers on mechanical properties of cellulose/epoxy composites", Composites: Part B 51 (2013), pp: 28–34. DOI: 10.1016/j.compositesb.2013.02.031

[25] Jawaid, M., and H. P. S. Abdul Khalil, "Cellulosic/synthetic fiber reinforced polymer hybrid composites: A review", Carbohydrate Polymers 86 (2011), pp: 1–18. DOI: 10.1016/j.carbpol.2011.04.043.

IOP Conf. Series: Materials Science and Engineering **574** (2019) 012014 doi:10.1088/1757-899X/574/1/012014

Performance Analysis of Abrasive Water Jet Cutting Process in Carbon Fiber Epoxy Polymer Composite

Binduk Potom[1], S. Madhu[2 *],S.Kannan[3], P. Prathap[4]

[1,2] Department of Automobile Engineering, Saveetha School of Engineering, Thandalam, Chennai, Tamilnadu, India
[3]Department of Mechanical Engineering, Hindustan College of Engineering and Technology, Coimbatore, Tamilnadu, India
[4] Department of Mechanical Engineering, Sri Krishna College of Technology, Coimbatore, Tamilnadu, India

Corresponding Author: mathumarine@gmail.com

Abstract. Abrasive water jet machining is unconventional material removal process used to dissolve openings and cavities by the force of rough particles of the slurry on hard and materials. It is a non traditional machining process able to machine wide scope of hard to cut materials such as alloys, composite materials. This work explores the impact of different process parameters on kerf angle and amount of materials removed which are imperative execution measures in abrasive waterjet machining. The variable abrasive water jet parameters considered here include water pressure, nozzle diameter stand-off distance and abrasive flow rate. Experiments were conducted on carbon fibre reinforced polymer composites by varying these parameters.This work concluded that maximum abrasive jet pressure increase the material removal rate and minimum stand off distance (SOD) will decrease the kerf angle in carbon fibre composites.

1. Introduction

Abrasive water jet(AWJ) cutting is a unconventionsalcutting procedure that utilizes high pressure water for delivering high speed stream, entrained with grating particles for a wide assortment of materials extending from delicate to hard materials. It is an adaptable procedure that can be utilized in numerous producing applications, for example, cutting, processing, cleaning, and surface treatment [1]. It has kind advantages like negligible heat affected zone in cutting procedure, high degree of accesability, and less machining time. In Abrasive water jet process, materials get removed by because of effect of high speed stream (blend of water and rough abrasive particles) on the surface to be machined [2]. The abrasive water jetmachining is fundamentally influenced by the different process parameters such as (i) Water jet pressure (ii) Abrasive flow rate (iii) Stand-off distance (iv) Abrasive jet traverse rate and (v) Diameter of nozzle[3]. In which kerf angle is a unwanted in abrasive water jet. The increased pressure and minimum stand-off distance decrease the kerf angle [4]. Material removal rate in directly influence by waterjet pressure. Also minimum stand-off distance (SOD) increases the mateiral removal rate [5].The quality of the machine surface depends upon the abrasive waterjet parameters. For improve the quality of the surfacedetailed knowledge of the effect of process parameters is required. In this way, to explore the impact of procedure factors on performance of AWJM, following parameters are considered in this paper [6]. In this work CFRP composite material is machined by abrasive water jet machining. Abrasive water jet pressure, nozzle diameter, flow rate and SOD were selected [7] as input process parmeters in this investigation. The mateiral removal rate (MRR) and Kerf angle were invetigated.

2. Materials and Method

2.1 CFRP Composite

In this work CFRP composites with 5 mm thickness were fabricated by hand layup process [8,9]. Epoxy rein was used as matrix material[10]. Hardner LY565 is used for mixing epoxy resin and carbon fibre. Unidirectional carbon fibre was used for this investigation.

2.2 Abrasive waterjet machining

Abrasive water jet is used for macgining CFRP samples. Holes were made on CFRP samples using the following waterjet parameters. The water jet pressure (2500 to 4000 bar) is used for machining the composite samples.The flow rate of the water along with abrasive particles is a major process parameter that affects the material removal rate in abrasive jet machining.It is the diameter of the tip of the nozzle [11,12]. It has a major impact on material removal rate of the composite.It is the distance between the nozzle tip and the workpiece. It influence of water jet parameters on material removal rate, diameter of hole and kerf angle were investigated. Kerf angle is a quantity which is often used to reflect the inclination of the kerf wall from the top surface kerf to the bottom of the kerf. The distance between the two kerfs determines the kerf angle of the material.

3. EXPERIMENTATION

Abrasive water jet investigations were done on CFRP composie as shown in Figure 1. Carbon fiber reinforced polymer was used as workpice material. CFRPare lightweight, solid materials utilized in the manufacturing of various items utilized in our day by day life. It is utilized in the modern, structural, and transportation fields [13, 14]. Abrasive water jet machining, which does not make a significantheat influenced zone, is more usable for machining of CFRP for present day applications[15].

Abrasive water jet parameters significanly affect the quality of the machined surface are considered in this work are pressure, flow rate, nozzle diameter and stand-off distance. The experiments were conducted on CFRP samples as shown in the Table 1.

Figure 1 Abrasive water jet machine

Table 1: Experimental parameters

S.No	Parameter	Unit	Levels		
			1	**2**	**3**
1	Pressure	bar	1500	2000	2500
2	Flow rate	gm/min	200	400	600
3	Nozzle diameter	mm	1.1	1.5	1.9
4	Stand-off distance	mm	3	5	7

The weight of the CFRP samples was measured using digital weighing machine before and after hole. The the top and bottom holes diameter was measured by using optical microscope. By using the material Plus software bottom and top diameter of the hole were eveluatedas shown in the Figure 3 and Figure 4.By using the equation 1 kerf angles were calculated and recorded. Totally 9 experiments were carried out for different abrasive waterjet parameters as ashown in Table 1. For each experuimets MRR and kerf angle were recorded as shown in Table 2

Table 2: Abrasive water Experimental results

Exp No	Input Parameters				Output Responses		
	P (Bar)	**FR (g/min)**	**SOD (mm)**	**ND (mm)**	**MRR (g/min)**	**TK (mm)**	**BK (mm)**
1	1500	200	3	1.1	5.8	2.202	2.067
2	1500	400	5	1.5	5.1	2.109	2.012

3	1500	600	7	1.9	6.0	2.538	2.228
4	2000	200	3	1.9	5.6	2.445	2.207
5	2000	400	5	1.1	6.2	2.674	2.319
6	2000	600	7	1.5	6.3	2.051	1.883
7	2500	200	3	1.5	6.4	2.656	2.193
8	2500	400	5	1.9	6.7	2.034	2.000
9	2500	600	7	1.1	6.5	2.496	2.252

4. Result and discussion

After conducting the waterjet experiments, the values of output parameters such as material removal rate (MRR), top kerf (TK) and bottom kerf (BK) were recorded in the Table 2.To investigate the quality of machined surface of CFRP composites the effect of abrasive water jet parameters on MRR, TK and BK were studied in this investigation. From this investigation it was seen that abrasive water pressure and SOD is the significant parameters [15, 8].

4.1 Effect of abrasive water jet parameters on MRR

From the figure it was seen that maximum water jet pressure increase the material removal rate in CFRP composite. When the pressure increases the kinetic energy of the abrasive particle increases which in turn erode the material surface rapidly. In this investigation maximum of 6.7 g/min MRR obtained.

From the experimental results it was seen that minimum stand-off distance (SOD) increase the material removal rate [8, 17]. When the distance between the work surface and the nozzle is minimum the high pressure abrasive particle coming out from the nozzle remove the material drastically.

4.2 Effect of abrasive water jet parametrs on Top and bottom kerf

In this current investigation top hole diameter depend upon the water jet pressure. When the pressure increases the kerf also increased as shown in Figure 2a. When the distance between the nozzle and the work surface is minimum [18], more amounts of abrasive particles impinging the work surface. Hence the top kerf increased.

When the pressure decreased, the abrasive particls reach the bottom surface with low kinetic energy.Hence the bottom kerf is smaller than the top kerf as shown in Figure 2b. When the stand-off distance (SOD) is minimum, the abrasive particles reach the bottom surface with more energy which in turn removes the material in CFRP composite material.

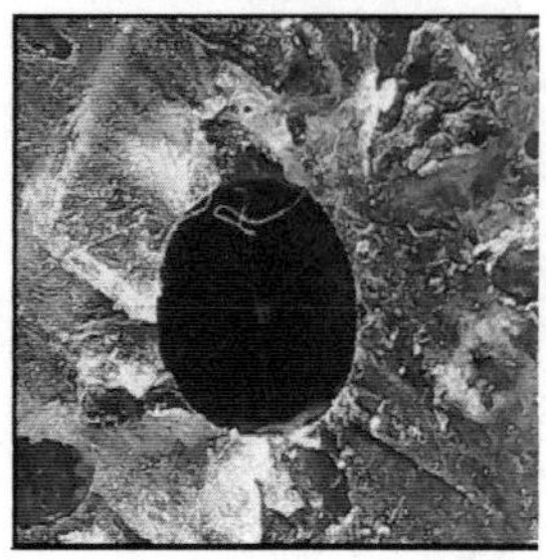

Top hole diameter

Bottom hole diameter

Figure 2 Microscopic images of top and bottom kerf of CFRP samples

5. CONCLUSION

From this investigation, it was obseved that abrasive water jet pressure and SOD are the significant parameters which affect the material removal rate and kerf charecterisitcs of the CFRP samples. The effect of water pressure and stand-off distance on kerf charecterisitics and MRR were studied. In this experimental research the following conclusion were drawn.When the water jet pressure is maximum, the MRR in the CFRP samples increased up to 6.7 g/min.The outcome of the research demonstrated that the top and bottom kerf varies according to the variable parameters provided, stand-off distance and abrasive flow rate.

REFERENCES

[1]Chen L, Siores E, Wong WCK. Optimising Abrasive Waterjet Cutting of Ceramic Materials. J Mater Process Technol. 1998; 74: 251–4p.

[2]Shimizu Z, Shinoda Y, Peng G. Flow Characteristics of Water Jet Issuing from a Fan Jet Nozzle. In Proceedings of 19th International Conference on Water Jetting; 2008 Oct 15–17; Nottingham, UK: 2008; 55–66p.

[3]Srinivasu DS, Axinte DA, Shipway PH, et al. Influence of Kinematic Operating Parameters on Kerf Geometry in Abrasive Waterjet Machining of Silicon Carbide Ceramics. Int J Mach Tools Manuf. 2009; 49: 1077–88p.

[4]M. A. Azmir, A.K. Ahsan, A. Rahmah, M.M. Noor, and A.A. Aziz, “Optimization of Abrasive Waterjet Machining Process Parameters Using Orthogonal Array With Grey Relational Analysis”,

Regional Conference on Engineering Mathematics, Mechanics, Manufacturing & Architecture (EM3ARC) (2007), pp 21~30.

[5] M. A. Azmirl, A.K. Ahsan, A. Rahmah, "Investigation on Abrasive Waterjet Machining of Kevlar Reinforced Phenolic Composite Using Taguchi Approach", Proceedings of the International Conference on Mechanical Engineering 2007 (ICME2007) (2007), pp 29-31.

[6] Conner I., Hashish M., Ramulu M. "Abrasive water jet machining of aerospace structural sheet and thin plate materials", in Proceedings of the 12th American Water jet Conference, Houston, USA, 2003.

[7] S Madhu, M Balasubramanian (2017), "Effect of abrasive jet process parameters on machining glass fibre reinforced polymer composite" Materials Science & Engineering Technology, 48 (11), pp 1146-1157.

[8] S.Madhu, M. Balasubramanian (2018) "Effect of swirling abrasives induced by a novel threaded nozzle in machining of CFRP composites", The International Journal of Advanced Manufacturing Technology Volume 95, 9-12, pp 4175-4189.

[9] Jeyaprakash P, Prabakaran V and Devaraju A. "Experimental and Numerical Analysis of Carbon Epoxy Fibre Composite under Buckling Load", *Materials Today: Proceedings, Vol. 5 (*2018), pp.14526-14530.

[10] Devaraju, A, Babu, K and Gnanavel babu, A . "Investigation on the Mechanical properties of Coconut Bunch fiber Reinforced Epoxy with Al_2O_3 Nano particles Composites for Structural Application", *Materials Today: Proceedings, Vol. 5 (*2018), pp.14252-14257.

[11] Khan A.A. and Haque M.M. "Performance of different abrasive materials during abrasive water jet machining of glass", Journal of Materials Processing Technology, Vol. 191, 2007.

[12] Jain N.K., Jain V.K. and Deb K. "Optimization of process parameters of mechanical type advanced machining processes using genetic algorithms", International Journal of Machine Tools & Manufacture, 47, 2007.

[13] M. Joseph Davidson, K. Balasubramanian, G. R. N. Tagore "Surface roughness prediction of flow formed AA6061 alloy by design of experiments" journal of Material processing Technology, vol 202,NIT Warangal,2008.

[14] Siddiqui T.U. and Shukla M. "Robust Parameter Design for Multi-characteristic Optimization of Abrasive Waterjet Cutting of Aramid Composite", Journal of Modern Manufacturing Technology, Vol. 1, 2008.

[15] Deaconescu A, Deaconescu T. Performance of Machining by Waterjet Erosion. Proceedings of the 1st International Conference on Manufacturing Engineering, Quality and Production Systems (Volume II). 2009; 2(1): 313–6p.

[16] John RozarioJegaraj J., Ramesh Babu N. A Soft Computing Approach for Controlling the Quality of Cut with Abrasive Waterjet Cutting System Experiencing Orifice and Focusing Tube Wear. J Mater Process Technol. 2007; 185(1–3): 217–27p.

[17] Shanmugam DK, Wang J, Liu H. Minimization of Kerf Tapers in Abrasive Waterjet Machining of Alumina Ceramics Using a Compensation Technique. Int J Mach Tools Manuf. 2008; 48: 1527–34p.

[18] Jankovic P, Igic T, Nikodijevic D. Process Parameters Effect on Material Removal Mechanism and Cut Quality of Abrasive Water Jet Machining. TheoretAppl Mech. 2013; 40(2): 277–91p.

The effect of palm biodiesel fuel on the performance of automotive four stroke diesel engine

T.Sivakrishna[1],S.Madhu[2], K.Sivakumar[3]

[1,2]DepartmentofAutomobileEngineering,SaveethaSchoolofEngineering,Thandalla m,Chennai,TamilNadu,India
[3]Department of Mechanical Engineering, Saveetha School of Engineering, Thandallam, Chennai, Tamil Nadu, India

Corresponding author: mathumarine@gmail.com

AbstractThe usage of biodiesel is been increased due to the shortage and cost inincrease of normal fuel in the market. The present investigation aims to explore the performance assessment of four stroke diesel engine using palm biodiesel.The palm biodiesel cannot be directly used as an alternative fuel so that the diesel and biodiesel is mixed with a concentration of (B10,D90), (B20,D80), (B30,D70), (B40,D60), (B100) and experimental investigation is experimental carried out on four stroke, double cylinder, to study engine performance.

1. INTRODUCTION

Palm biodiesel is created from the consumable vegetable oil from the product of the oil palm tree. The palm organic product is the well spring of both palm oil(removed from palm natural product) and palm portion oil(separated from the organic product seeds).Palm oil itself is ruddy since it contains a high measure of beta carotene. Palmoil's development in the market comes along while after the presentation of ethanol, produced using sugarcane, and different added substances. The addition intrigue and utilization of palm biodiesel have brought about expanding concern on environmental effect of palm tree development just assustenance versus fuel dilemma. The particular ranger service says from expanded palm biodiesel generation stays uncertain. This is on the grounds that the spread of palm tree ranches is forthemost part because of the expanding nourishment and industry demands. There are various favourable circumstances in utilizing palmoil for the generation of biofuel. In contrast to petroleum derivatives, the ignition of palmoil biofuel does not expand the dimension of carbondioxide in the climate as the oil is only returning carbondioxide gotten before from the air through photosynthesis, biofuel is viewed a scarb on neutral. Since carbondioxide is an ozone-depleting substance incharge of a dangerous atmospheric deviation, the advantages by the consuming of biofuel rather than a non-renewable energy source. This palm biodiesel is been blended with biodiesel at various blend ratios and been used to test the engine performance. The bio-diesel fuels were produced from palmoil using the trans esterification process with lowmolecular weight alcohols and sodium hydroxide then tested on a steady state engine testrig using a 4 four cylinder Compression Ignition(CI) engine. This study also show by blending bio-diesel with diesel fuel at intervals of B10,B20,B30, B40 and B100 c and e crease harmful gas emissions significantly while maintaining similar performance output and efficiency.

Production optimization was achieved by changing the variables.Several research group shave investigated the properties of a bio-diesel blend with soybeanoil methyl esters in diesel engines and found that particulate matter CO, and so to mass emissions decreased, while NOx increased, examined the performance and exhaust emissions of rapeseedoil methylesters indirect injection diesel engines, and found that there were lower emissions of CO, CO2 and HC. Similar results were form ethyl esters of sunflower oil and palmoil when they were blended with marine diesel and tested in a stationary diesel engine. Studied the fuel properties of palm esters blended with diesel from 20% to 80% by volume. It was found that B20 (a blend of 20% bio-diesel and 80% petroleum diesel) a could be used as an appropriate alternative fuel to petroleum diesels because they apparently produced less CO, NOx emissions, and smoke density, confirmed that emission of polycyclic aromatic hydrocarbons(PAH) decreased when the ratio of palm. The results showed improved engine performance and reduced exhaust gas emissions with levels acceptable to the standard.

2. Materials and Method

2.1 Preparation of fuel blend

The tranesterified plam biodiesl has been collected seperately. Palm biodiesel and the fuel mixture is mixed with the ratio such as (10% biodiesel + 90% diesel) , (20%biodiesel+80%diesel),(30%biodiesel+70%diesel)(40% biodiesl+60diesel),(100% biodiesel) and is mixed and titrated.

Step1: 100% of pure Palm biodiesel.
Step2:100% of pure diesel.
Step3:The fuels are blended in different ratios B10,B20,B30,B40,B100.
Step4:These fluid blends are testing in two cylinder diesel engine to find out the Performance combustion and emission.

Fig1: palm biodiesel

Fig2: diesel

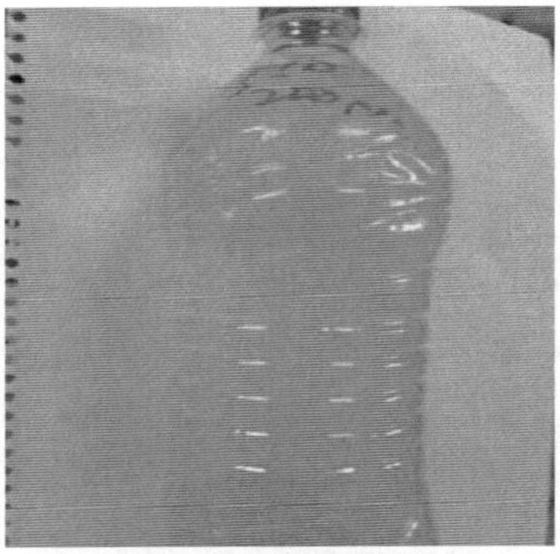

Fig3 blended diesel

3. EXPERIMENTAL SETUP

For this current investigation, Simpsons make two barrel four stroke motor engine is used. This engine present-day innovation framework it is utilized for research work and fuel test it creates motor test in an elective fuel which begins with yield control at 21 Kw and consistent speed at 2000 rpm and it has a motor limit 1670 cc the activity condition may vary for motor setup for this production framework it has pressure proportion is 18 5:1 .the broke and stroke 127mm X 2mmalong with pole get together the motor particular is referenced Table 1 and two chamber four stroke motor. The motor stacking is an electric dynamometer which has a variety in burden control it relies upon the motor lower to higher burden from these examinations were performed diverse burdens 25ppm.50ppm.75ppm,100ppm separately the air cooling test motor utilized water and liter dependent on burdens and the fuel siphon.

Table 1 Engine specification used for this investigation

Parameters	Specification
Engine type	Four stroke two cylinder engine
Manufacture	Simpsons Ltd
Load	Electric dynamometer
Rated power	28 bhp@ 2000rpm
Fuel pump	Mico bosch
Bore and stroke	127 mm X 2 mm
Cylinder Capacity	1670 cc
Compression ratio	18 5:1
Cooling system	Water
Electrical system	12 Volts (dynamo/alternator)
Flywheel	SAE 1
Engine starting system	Electrical

IOP Conf. Series: Materials Science and Engineering **574** (2019) 012015 doi:10.1088/1757-899X/574/1/012015

Figure 4 Engine Test rig

4. RESULT AND DISCUSSION

4.1 Effect of load on CO

Experimental result demonstrated that with utilizing paln biodiesel blends,the motor power and output torque increment while brake explicit fuel utilization decline primary reason improve the CI motor execution attributes when contrasted and that of flawless diesel fuel because of the higher oxygen substance of the mixed energies that have improved the diffusion burning stage and diminished the ignition time.

The above shown Fig5 shows the result of the five blends CO emission percentage and from the graph it describes that the diesel and other blends b10,b30,b40,b100 emits more CO than the b20 blend from the above shown fig the b20 emiits lesser CO than other blends.

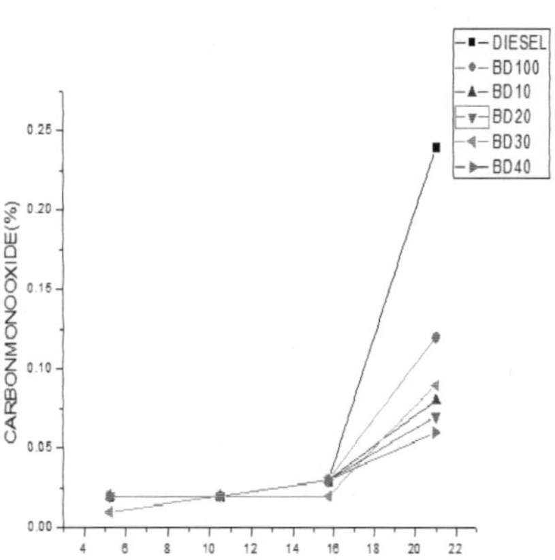

Fig 5 Carbon monoxide (CO)

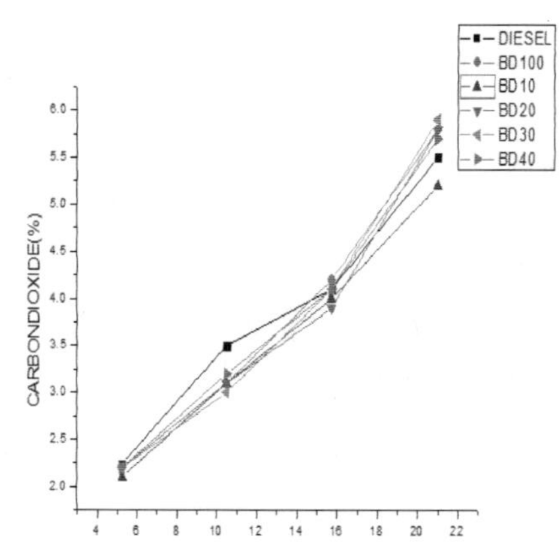

Fig6 CO2

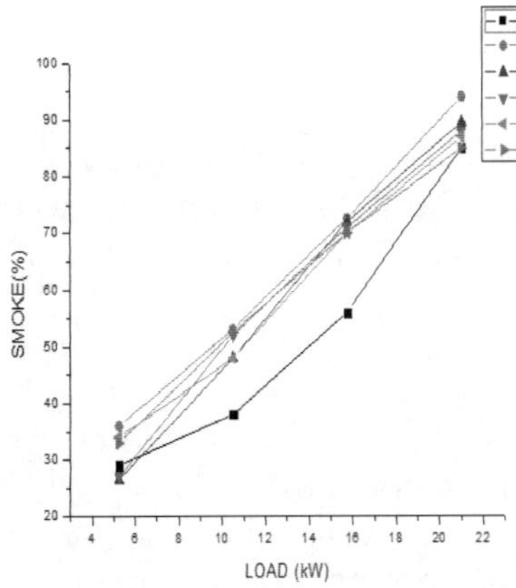

Fig7 Smoke

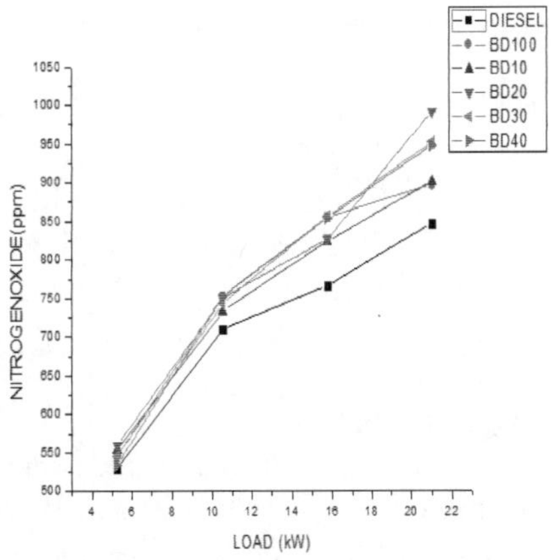

Fig8 Nox

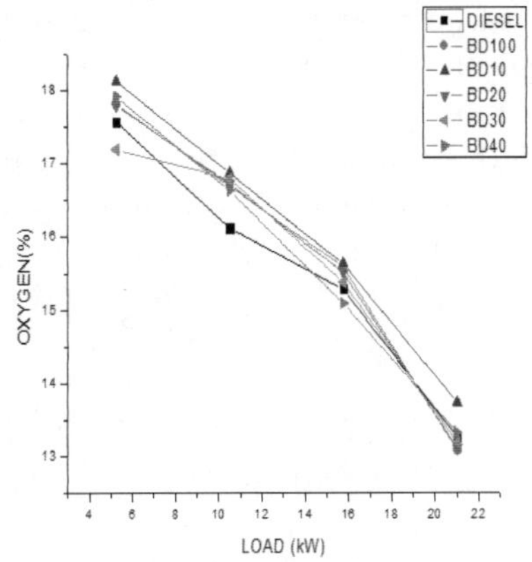

Fig9 Oxygen

4.2 Effect of load on CO_2

The above showing Fig6 dicribes the results of CO2 emitted by the blends tested in the engine and the b30, b40, b100 emittes more CO2 than b10, b20 blends and the b10 emmits lesser co2 than the b20 blend but there is no larger difference between the two blends.

4.3 Effect of load on smoke

The result from the above fig7 describes the smoke emitted from the blend series and the higher smoke produced is from the blend b100,followed by the b10,b30,b40 and the most lesser smoke is produced by the pure diesel and followed by the b20 blend.

4.4 Effect of load on NOx

The Fig8 is the results of the blends emiiting Nox and the higher percentage of Nox is emitted from the blend b30 constantly though the b20 is the blend which produce the five gases lesser than the other blends when the engine is runned at higher rpm the b20 emiits more Nox than the other gases and the b10 produces lesser Nox than other blends at all RPM and followed by te normal pure Diesel.

4.5 Effect of load on O_2

The emission of O_2 content should be higher in every fuel where the above shown fig9 discribes the amount of oxygen emitted from the fuel blends series and the B10, b40, b20 blends emitts more oxygen than the other blends b30,b100,diesel which proves that the b20 is an satisfying ratio which can be used to blend many type of bio fuels.

5. CONCLUSION

The investigation of the undertaking manages the mix of biodiesel, and diesel. At first, the diesel motor was kept running with crude diesel 100%and biodiesel(Palm biodiesel) 100% which is acquired from fuelwith blend are B20 consequently it was stacked in diesel motor with 1200 rpm speed. it was discovered the outflow normal for hydrocarbon ,carbon dioxide, carbon monoxide, nitrogen oxygen and smoke it demonstrates the variety of discharge alongside impact of palm oil indicated constructive outcome on variety of emanation over all the test energizes among the distinctive test fuel B20 are the best it keep up steady and consistent to produce less measure of discharge . In higher surface region the volume proportion of outflow was less creation at greatest burden was found in B20 demonstrated the best result when contrasted and others .amid variety of smoke the demonstrated the best outcome on less emanation it control the smoke generation and lessens the earth contamination by utilizing fuel with nano additive .thus it was at long last inferred that the utilizing of palm biodiesel with diesel mixes decreases the outflow and give better ability, execution in the diesel motor.

REFERENCES

[1] Rolvin D'Silva, Binu K.G, Thirumaleshwara Bhat Performance and Emission characteristics of a C.I. Engine fuelled with diesel and biodiesel fuel additive (2015).

[2] L.Jeryrajkumar, G.Anbarasu, T.Elangovan Effects on Nano Additives on Performance and EmissionCharacteristics of Calophyllim inophyllum Biodiesel (2016).

[3] Biogas Potential from Vetiveria zizaniodes (L.) Planted for Ecological Restoration in China, Yuying Lia, Xuemin Ren, Erik Dahlquist, Panpan Fana, Tang Chaoa.

[4] H.H. Masjuki, M.A. Kalam, M. Syazly, T.M.I Mahlia., A.H. Rahman, M. Redzuan, M. Varman.

[5] R. Saidur, and Y.H. Yau Tribology and Engine Testing Laboratory (2015)177–188.

[6] Numerical Simulation for Effect of Fuel Injector Parameters on Combustion and NOx Emissions of STC Diesel Engine Xinguang Li, Yinyan Wang and Yang Wang.

[7] Study on Combustion and NOx Emission Characteristics of a DI Diesel Engine Operating on Pistache Seed Biodiesel/Diesel and Methanol/Diesel Blends Chenyang Fan Bin XuZhihao MaJian Wu.

[8] Experimental study of a HD Diesel Engine Equipped with Urea-SCR System Able to Achieve the Euro IV Emission Limits,Chenglin DENG ,Automotive Safety and Energy

National Laboratory Tsinghua University Beijing, China ,Liping HUANG, Hailong PANG, Xinyun ZI, Hao LI, Jinyong HE Power Equipment Energy-saving and Emission-control.

[9] Jagannath Balasaheb Hirkude & Atul S.Padalkar 2012, 'Performance and emission analysis of a compressionignition Engine operated on waste fried oil methyl esters', Applied Energy.

[10] Costa R. E. & Lora, E.E.S., "The energy balance in the production of palm oil biodiesel - two case studies.

[11] Demirbas, A., "Progress and recent trends in biofuels", Progress in Energy and Combustion Science 33 (2007), pp. 1–18.

[12] Fernando, S., Hal, C. and Jha S., "NOx Reduction from Biodiesel Fuels", Energy & Fuels 2006, 20, pp. 376-382.

[13] Friedrich, S., "A world wide review of the commercial production of biodiesel - A technological, economic and ecological investigation based on case studies", Des Institutes für Technologie und Nachhaltiges Produktmanagement, Vienna, 2004.

[14] Frondel , M., Peters, J., "Biodiesel: A new Oildorado?", Energy Policy 35, pp. 1675–1684.

[15] Pinto, "Biodiesel: An Overview", Journal of the Brazilian Chemistry Society., Vol. 16, No. 6B, 2005, pp. 1313-1330.

[16] Sheehan, J., "Overview of Biodiesel and Petroleum Diesel Life Cycles", National Renewable Energy Laboratory, USA, Colorado, May 1998.

IOP Conf. Series: Materials Science and Engineering **574** (2019) 012016 doi:10.1088/1757-899X/574/1/012016

Effect of Tio2 Nanoparticles In Spirulina Biodiesel- Diesel Blends at Various Injection Strategies: Performance, Combustion and Emission characteristics

Suresh yaramati[1], Sakthimurugan.V[2]

1.U.G scholar department of automobile engineering saveetha school ofengineering Saveetha university Chennai –602105
2.Assistant professor department of automobile engineering saveetha school ofengineering Saveetha university Chennai –602105

Abstract. Since so many years all the automobile engines are running by using the fossil fuels. Day by day the usage is increasing. So, by using the fossil fuels it reaching the last stage and finally it leads to the depletion. So, the fossil fuels are the responsible for air pollution, global warming, and the gradually increasing of global climate. So, the algae biofuels are considered to be the most renewable source of energy and also it is eco-friendly. Microalgae considered to be the potential source which can produce biofuel from algae. Nitrates, phosphates, and carbon dioxide which were produced from atmosphere are consider to be growing environment for algae. Spirulina is a blue-green algae. Autotrophic and alkallophyletic are the two attributes of spirulina. For spirulina lesser amount of sunlight is needed than the homes. pH is the important ingredient in the culture medium. After the transesterification process about 87.75% of biodiesel yield was obtained at optimum level conditions. By comparing of both spirulina bio-oil and chlorella bio-oil the parameters like viscosity and density of chlorella were very equally to the conventional diesel fuel. By the above reasons spirulina is strongly considered to be a potential raw material or feedstock for the biodiesel production to be used in diesel engines. For this investigation to measure the engine performance parameters like exhaust temperature, brake specific fuel consumption and emission parameters like hydrocarbons, carbon dioxide, nitrogen oxides, oxygen, carbon monoxides and smoke are measured by changing the injection timings like retard 23 degree, retard 21 degree and retard 19 degree. Finally B20 is the blend taken and this blend is mixed with 75 ppm of Titanium dioxide as a nanoparticle experiment was conducted on four stroke, twin cylinder diesel engineused.

1.Introduction

Spirulina based biodiesel is a long chain fatty acids made up of mono-alkyl esters (1). It is beneficial to the environment because it produces lesser emissions (2, 3). Due to the less amount of yielding of oil leading to the more amount of feedstock is required for the production of biodiesel (4). The larger amount of usage of fossil fuels is lead to the global warming. To reduce these consequences the scientists are searching the new technique to yield more amount of oil in less land. Biofuel is an energy contained fuel from living organism. Biomass is the raw material for the production of biofuel. The convenient energy contains in the three varieties of ways such as thermal conversion, chemical conversion and biochemical conversion (5). Pure form of ethanol is used to run the vehicles. To increase octane number and to reduce the engine emissions it is used as gasoline additive. The production of biodiesel from food crops creating a new competition to the human food. So the world is adopting algae biodiesel. Nuclear energy sources are coming for extinction. The wind energy and solar energy don't have the capability of reaching the requirements for alternative energy. Due to the poor technological developments it is difficult to use wind, solar and other forms of energy to replace the fossil fuels. Here some of the advantages are listed than petrol and diesel fuels(6).

Biodiesel can be used as a fuel in present engines without any engine modifications.

- Vegetables are the main source of biodiesel so it does not contain any metal or crude residues, Sulphur and aromatic hydrocarbons.
- Occupational safety and health administration states that it is a non flammable liquid(7).

Future is ready to adopting the perfect option is to selecting biodiesel (8). Esterifies vegetable oils are known to be biodiesel, non-edible vegetable oils are called as straight vegetable oils, ethyl esters or methyl esters are to be considered vegetable oils (9). Rural areas are the best place to gain the vegetable oils easily they are with high cetane number so they can blend easily with diesel (10-13). (14) states that sunflower oil, rapeseed oil, jatropha oil, coconut oil, algae oil can be used in IC engines. (15) states that microalgae fuel properties were evaluated. The research proved that waste cooking oil biodiesel and diesel blends are good alternative fuel for diesel engines (16, 17). (18) has stated that waste cooking oil biodiesel are good alternative fuel. The results which were obtained by the experiment on influence of soybean biodiesel on DI diesel engine performance, emission and combustion characteristics are to be known that biodiesel is an eco-friendly fuel (19). (20) conducted an experiment by using the sources like pongamia and jatropha and conducted that biodiesel can be used as an alternative fuel in the near future. (21) done an experiment with mahuaoil conducted that mahuaoil ethyl ester can be used as a substitute. (22) studies the minimization of VCR-CI fueled with dunaliellasalina oil and corn oil using taguchi approach finalized that CO was lower than diesel. (23) NOx emissions are increased by increasing of the combustion temperatures known by conducting an experiment named investigated the effects of preheated cotton seed oil methyl ester on performance and emission of diesel engines. (24) accepted that pine oil can be used as a fuel by without any trans esterification process. (25) states that by conducting an experiment impact of compression ratio and injection parameters on performance and emission of DI diesel engine states that by increasing of compression ratio then temperature is also reaches maximum. At higher compression ratio lesser emission will produced from biodiesel (26). (27) studied the effects of compression ratio on an engine filled with waste oil finally concluded that biodiesel has more benefits than pure diesel. (28) analyzed the usage of chicken fat oil blended with diesel CI engine. Generally the microalgae is more efficient because it convert into biomass by taking lesser amount of energy and CO_2 (29). Algae have the good characteristics is that they grow in lands which are not fit for food (30). Microalgae produces 40 times more amount of fuel per acre than plants (31). Other than biodiesel a lot of products can make such as pharmaceutical and cosmetic products (32). Due to the presence of phyto nutrient value and pigments can be used in therapeutics and diagnostics (33, 34). Algae has have the capacity to cure some of the diseases like anemia, tumor growth malnutrition (35, 36). Algae can be served as fish food and animal food (37). Adjustment of engine injection tuning is possible by based on blend level to improve engine emission and performance (38). Spirulina is a valuable product having high nutrition value (39-42). It is the major source for B_{12} vitamin (43). Absence of cellulose in its cell wall 85- 95% is along with an organism (44). Arthrospira plantesis is available in tropical regions with the concentrations of Nacl and bicarbonatesare the concentrations available in the tropical regions of arthro spira plantesis (45-48).

2. Materials and methodology

2.1 Blending of biodiesel with diesel

Actually the blending can be defined as the mixing or adding the things together to form like one thing. In fuels purpose the biodiesel is mixing with diesel. Splash method and inline methods are the two methods of blending. The blending work has done correctly in the proportions by using fuel measuring instruments. B20 is only one blend taken for the research purpose and it contains 80% of diesel fuel and 20% of spirulina biodiesel finally this B20 is mixed with 75 ppm of titanium dioxide (Nanoparticle).

2.2 Experimental setup

The engine setup is explained through the help of a schematic diagram by figure 1. Engine specifications are discussed in the table 1. For this research twin cylinder four stroke simpson diesel engine is selected. The engine was connected with dynamometer along with this computer system is present for knowing the readings. With the help of stopwatch measure the fuel consumption. The exhaust gases temperature are measured with thermocouple and with temperature indicator. U tube manometer is placed to air tank measure the air consumption. For measuring the exhaust emissions and smoke intensity gas analyzer and smoke meter are taken. At no load condition engine was run for 300 seconds by using diesel fuel. And then maintain the constant 1500 rpm. After the fuel consumption has completed several engine performance parameters and emissions are measured.

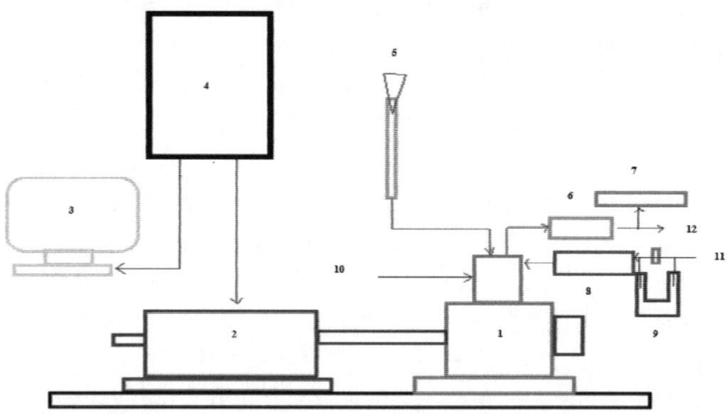

Figure 1: Schematic diagram of engine setup

Equipments

1. Engine
2. Dynamometer
3. Computer
4. Controlpanel
5. Burette
6. Gasanalyzer
7. Smokemeter
8. 8. Air tank orbox
9. U-tube manometer
10. Cooling water
11. Air
12. Exhaust gases

IOP Conf. Series: Materials Science and Engineering **574** (2019) 012016 doi:10.1088/1757-899X/574/1/012016

Manufacturer	Simpson
Type of engine	4 stroke CI engine
No. of cylinders	Twin
Compression ratio	17.5
Bore	91.40mm
Type of cooling	Water cooling
Rated power	24 BHP @ 2000 RPM
Loading device	Eddy current

Table 1 Engine specifications

3. Results and discussion

3.1 Brake Specific Fuel Consumption(BSFC)

The comparison of BSFC with load is plotted in the fig.2. From the fig.2 we are knowing that BSC is gradually reducing when there is an increasing of load. Retard 21 degree is showing lesser amount of BSFC. But retard 19 degree is showing the maximum amount of BSFC.

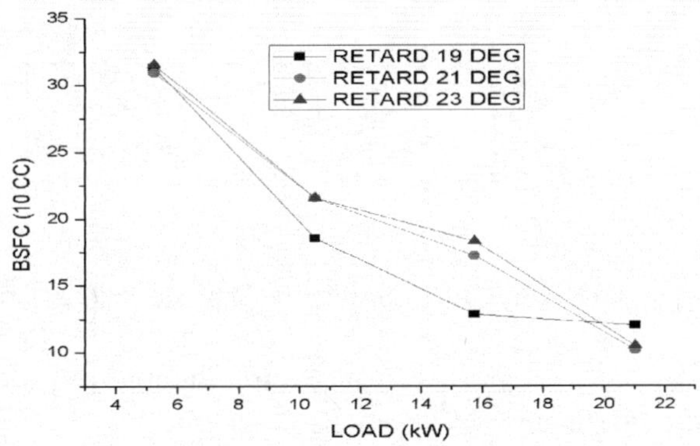

Fig.2 Load vs BSFC

3.2 Exhaust temperature

The variation of exhaust temperature with load is plotted in the fig.3. From the fig.3 knowing that exhaust temperature is gradually increasing by increasing of load. Retard 19 degree is possessing lesser temperature and maximum temperature at minimum load condition and in maximum load conditions. But retard 23 degree is showing more amount of exhaust temperature at both minimum and maximum load conditions.

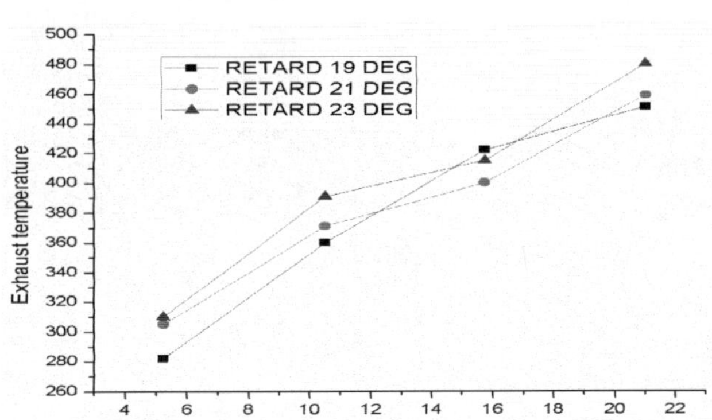

Fig.3 Load vs Exhaust temperature

3.3 Carbon monoxide(CO)

The comparison of CO with load is plotted in fig.4 from the fig.4 knowing that amount of CO is gradually increasing by the increasing of load conditions. All retards are at same point when the load condition is minimum. But at maximum load condition both retards 21 and 23 are at minimum level of CO but retard 19 degree is showing the maximum level of CO.

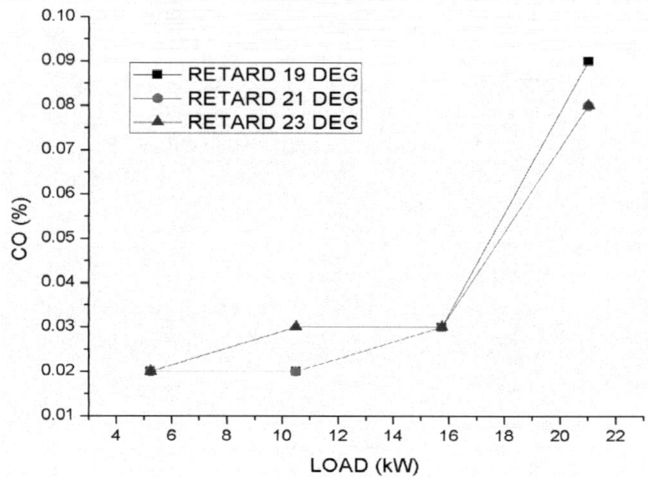

Fig.4 Load vs CO

3.4 Hydrocarbons(HC)

The variation of HC with load conditions are plotted in the fig.5. There is a great value of retard 19 degree of HC are observed at some instant condition of load. But finally it follows the remaining retards at both minimum and maximum load conditions. It is observing that all retards are pointed at same point at minimum and maximum load conditions.

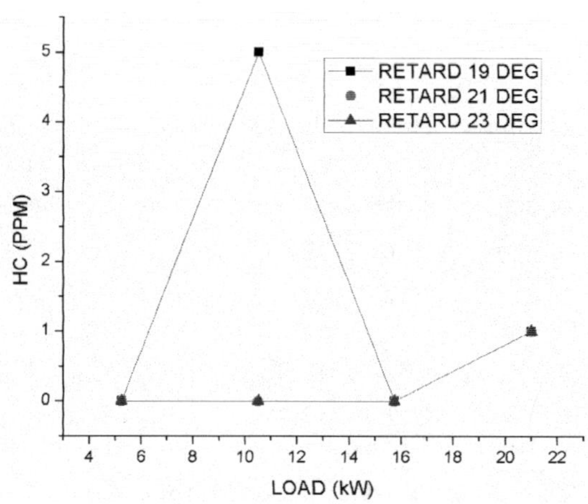

Fig.5 Load vs HC

3.5 Nitrogen oxides(NOx)

From the fig.6 knowing that the comparison of NOx with load. It is knowing that NOx is increased by the increasing of load conditions. Retard 21 degree is showing the lesser amount of NOx at both minimum and maximum load conditions. But retard 19 degree is showing the higher amount of NOx at minimum and maximum load conditions.

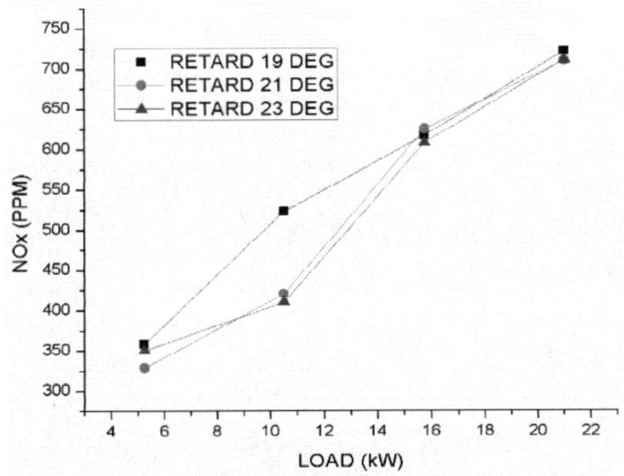

Fig.6 Load vs NOx

3.6 Oxygen(O2)

The comparison of O2 with the load is plotted in the fig.7. The amount of oxygen is gradually reduced by the increasing of load is known from fig.7. Retard 19 degree showing lesser amount of oxygen at minimum load condition while at maximum load condition it is showing more. At minimum load conditions both retard 21 and 23 degree are showing higher amount of oxygen. But retard 19 degree is showing higher amount of oxygen at maximum load conditions.

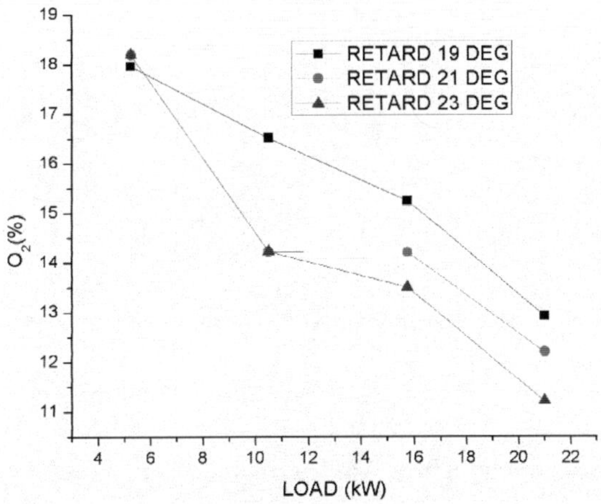

Fig.7 Load vs O2

3.7 Carbon dioxide(CO2)

The amount of CO2 is increasing by the increasing of load condition were known by the fig.8 and it is telling that retard 19 degree is showing lesser amount of CO2 an higher at minimum and maximum load conditions But in case of retard 21 degree is slightly more amount of CO2 at minimum and maximum loadconditions.

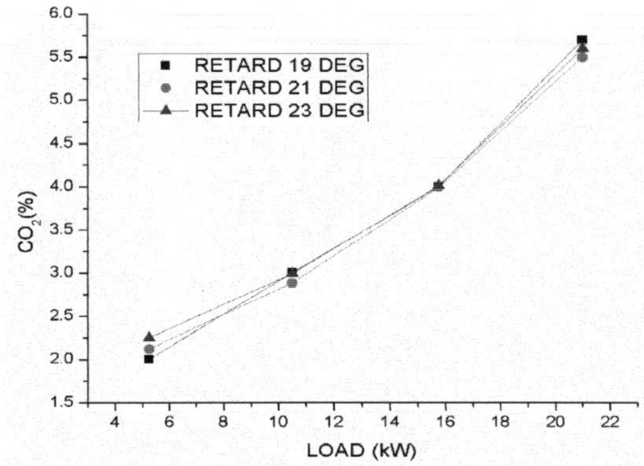

Fig.8 Load vs CO2

3.8 Smoke

The gradually increasing of smoke by the increasing of load is shown in fig.9. Retard 19 degree showing lesser amount of smoke and maximum amount at minimum and maximum load conditions. Incase of retard 21 degree is more amount of smoke and lesser amount of smoke is observed at minimum and maximum load conditions.

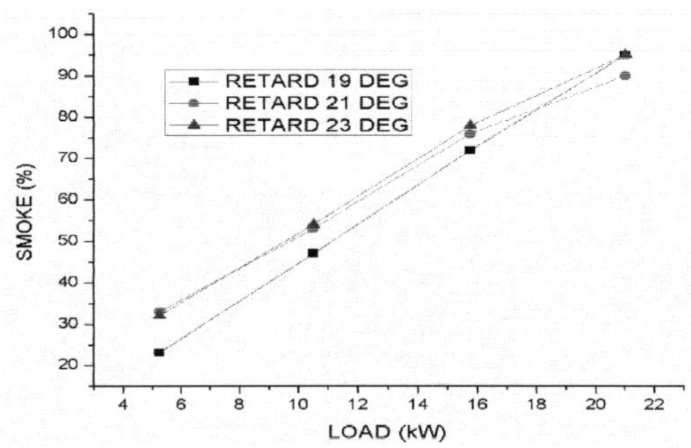

Fig.9 Load vs Smoke

4. Conclusion

This research is concluded that effects of biofuel in the engine performance and emission characteristics of diesel engine by using spirulina biodiesel. From this research retard 21 degree showing best results for brake specific fuel consumption, smoke and carbon dioxide. Retard 23 degree showing the best results for oxygen. And last but not the least retard 19 degree showing the best results for exhaust temperature, NOx, carbon monoxide and hydro carbons. From this research more best results are showing retard 19 degree.

5. References

[1] Knothe, G., Krahl, J., Van Gerpe, J., 2005. The Biodiesel Handbook. American Oil Chemist's society (AOCS) Press, Champagn,IL.
[2] Knothe, G., 2005. Dependence of biodiesel fuel properties on the structure of fatty acid alkyl esters. Fuel Process. Technol. 86,1059–1070.
[3] Lapuerta, M., Armas, O., Rodrı´guez-Ferna´ ndez, J., 2008. Effect of biodiesel fuels on diesel engine emissions. Prog. Energy Combust.Sci. 34,198–223.
[4] Mostafa, S.S. and El-Gendy, N.S., 2017. Evaluation of fuel properties for microalgae Spirulina platensis bio-diesel and its blends with Egyptian petro-diesel. *Arabian Journal of Chemistry*, *10*,pp.S2040-S2050.

[5] Bruhn, A., Dahl, J., Nielsen, H.B., Nikolaisen, L., Rasmus- sen, M.B., Markager, S., Olesen, B., Arias, Burgess, R M., Pelletier, M. C., Ho, KT.. Serbst, J.R., Ryba, SA., Kuhn, A., Perron, M. M., Raczelowski, Greer, D.E., &Ziebell, C.D. (1972). Biological removal of phosphates from water, Journal of the Water & Jensen, P.D. (2011). Bioenergy potential of Ulva lactuca: Bio- mass yield, methaneproduction an combustion, Biore- source Technology, 102,2595-260-4.

[6] The viscosity after the transesterification process was 4.47mm2/s at 400C which is 63.3% higher than DF. The viscosity of BD highly decreased after the transesterification process by70%.

[7] Rahman, M.A. and Nahar, K., 2016. Production and Characterization of Algal Biodiesel from Spirulina Maxima. *Global Journal of Research InEngineering*.

[8] Tashtoush, G. M., M. I. Al-Widyan, A. M. Albatayneh (2007). Factorial analysis of diesel engine performance using different types of biofuels. *Journal of Environmental Management* 84,401–411.

[9] Jindal, S., B. P. Nandwana, N. S. Rathore and V. Vashistha (2010). Experimental investigation of the effect of compression ratio and injection pressure in a direct injection diesel engine running on Jatropha methyl ester. *Journal of Applied Thermal Engineering* 30,442-448.

[10] Narayana Reddy, J. and A. Ramesh (2006). Parametric studies for improving the performance of a jatropha oil fueled compression ignition engine. *Renewable Energy* 31,1994-2016.

[11] Avinash, K. A. and A. Dhar (2013). Experimental investigations of performance, emission and combustion characteristics of Karanja oil blends fuelled DICI engine. *Renewable Energy* 52,283-291.

[12] Dhandapani, K. A., B. SenthilkumarPachamuthu, A. Md. NurunNabi, J. EinarHustad and T. Løvås (2012). Theoretical and experimental investigation ofdiesel engine performance, combustion and emissions analysis fueled with the blends of ethanol, diesel and jatropha methyl ester. *Energy Conversion and Management* 53; 322–331.

[13] Hariram, V., G. and M. Kumar (2013). Combustion Analysis of Algal Oil Methyl Ester in a Direct Injection Compression Ignition Engine. *Journal of Engineering Science and Technology* 8(1), 77 –92.

[14] Shehata, M. S. and S. M. Abdel Razek (2011). Experimental investigation of diesel engine performance and emission characteristics using jojoba/diesel blend and sunflower oil. *Journal of Fuel* 90;886–897.

[15] Soha, S. M. Mostafa, S. Nour, S. El-Gendy. Evaluation of Fuel Properties for Microalgae Spirulina Platensis Bio-diesel and its Blends with Egyptian Petro-diesel. *Arabian Journal ofChemistry.*

[16] Muralidharan, K., D. Vasudevan, K. N. Sheeba. (2011). Performance, emission and combustion characteristics of biodiesel fueled variable compression ratio engine. *Journal of Energy* 36;5385-5393.

[17] Wail, M. Adaileh and S. Khaled AlQdah (2012). Performance of Diesel Engine Fuelled by a Biodiesel Extracted from a Waste Cocking Oil. *Journal of Energy Procedia* 18;1317-1334.

[18] Ozer, C. (2014). Combustion characteristics, performance and exhaust emissions of a diesel engine fueled with a waste cooking oil biodiesel mixture. *Journal of Energy Conversion and Management* 87;676–686.

[19] OrkunOr, L Yuksek, A TekinErgenc and M. Ozkan (2014). Effects of soybean biodiesel on a DI diesel engine performance, emission and combustion characteristics. *Journal of Fuel* 115,875–883.

[20] Balajee, D., G. Sankaranarayanan, P. Harish, N. Jeevarathinam (2013). Performance and combustion characteristics of ci engine with variable compression ratio fuelled with pongamia and jatropha and its blends with diesel, *International Journal of Mechanical Engineering and Robotics Research* 2(3), ISSN 2278 –0149.

[21] SukumarPuhana, N., G. Vedaramana, Sankaranarayanana, V. Boppana Bharat Rama (2005). Performance and emission study of Mahua oil (madhucaindicaoil) ethyl

esterin a 4-stroke natural aspirated direct injection diesel engine. *Journal of Renewable Energy* 30(8),1269–1278.

[22] Balajee, D., G. Sankaranarayanan, K. Sai Prashanth (2015). Optimization of a stationary VCR–CI engine fuelled with Dunaliella Salina oil and Corn Oil using Taguchi approach *International Journal of Applied Engineering Research* 10(19). ISSN 0973-4562

[23] Murat, K., G. Ergen and M. Hosoz (2008). The effects of preheated cottonseed oil methyl ester on the performance and exhaust emissions of a diesel engine. *Journal of Applied Thermal Engineering* 28,2136-2143.

[24] Vallinayagam, R., S. Vedharaj, W. M. Yang, P. S. Lee, K. J. E. Chua, S. K. Chou (2013). Combustion performance and emission characteristics study of pine oil in a diesel engine. *Journal of Energy* 57;344-351.

[25] Cenk, S. and M. Gumus (2011). Impact of compression ratio and injection parameters on the performance and emissions of a DI diesel engine fueled with biodiesel-blended diesel fuel. *Applied Thermal Engineering* 31;3182-3188.

[26] Bhaskor, J., K. Bora, U. Saha, S. Chatterjee and V. Veer (2014). Effect of compression ratio on performance, combustion and emission characteristics of a dual fuel diesel engine run on raw biogas. *Journal of Energy Conversion and Management* 87;1000–1009.

[27] EL-Kassaby, M., A. Medhat and N. Allah. (2013). Studying the effect of compression ratio on an engine fueled with waste oil produced biodiesel/diesel fuel. *Alexandria Engineering Journal* 52,1–11.

[28] Lakshmanan, P., P. Kaliyappan, M. Ranjithkumar, K. Aravinth, D. Vakkachan, C. Moorthy and S. Kumar, (2017). An Experimental Investigation to Study the Performance and Emission Characteristics of Chicken Fat Oil Fuelled DI Diesel Engine. *Journal of Applied Fluid Mechanics* 10, Special Issue, pp.85-91.

[29] Sudhakar, K., Suresh, S., Premalatha, M., 2011. An overview of CO2 mitigation using algae cultivation technology. Int. J. Chem. Res. 3 (3),110–117.

[30] Khan, Sh., Rashmi, A., Hussain, M.Z., Prasad, S., Banerjee, U.C., 2009. Prospects of biodiesel production from microalgae in India. Renewable Sustainable Energy Rev. 13,2361–2372.

[31] Schenk, P.M., Thomas-Hall, S.R., Stephens, E., Marx, U.C., Mussgnug, J.H., Posten, C., Kruse, O., Hankamer, B., 2008. Second generation biofuels: high-efficiency microalgae for biodiesel production. Bioenerg. Res. 1,20–43.

[32] Scott, S.A., Davey, M.P., Dennis, J.S., Horst, I., Howe, C.J., Lea- Smith, D.J., Smith, A.G., 2010. Biodiesel from algae: challenge and prospects. Curr. Opin. Biotechnol. 21,277–286.

[33] Becker, E.W., 1994. Microalgae. Cambridge Univ. Press, Cambridge, NewYork.

[34] Vonshak, A., Tomaselli, L., 2000. Arthrospira (Spirulina) systematic and ecophysiology. In: Whitton, B.A., Potts, M. (Eds.), The Ecology of Cyanobacteria in the Netherlands. Kluwer AcademicPublishers.

[35] Hernandez-Corona, A., Nieves, I., Meckes, M., Chamorro, G., Barron, B.I., 2002. Antiviral activity of Spirulina maxima against herpes simplex virus type2. Antiviral Res. 56 (3),279–285.

[36] Mendes,R.F.,Nobre,B.P.,Cardoso,M.T.,Peveira,A.,Palavra,A.F.,2003.
   i. Supercritical carbon
   ii. dioxide extraction of compounds with pharmaceutical importance from microalgae. Inorg. Chim. Acta
   iii. 356, 328–334.

[37] Becker, E.W., 2004a. Microalgae in human and animal nutrition. In: Richmond, A. (Ed.), Handbook of Microalgae Culture. Biotechnology and Applied Phycology. Blackwell Science,Oxford.

[38] Tat, M.E., Van Gerpen, J.H., 2003. Biodiesel blend detection with a fuel composition sensor. Appl. Eng. Agric. 19 (2),125–131.

[39] Oliveira, M.A.C.L.; Monteiro, M.P.C.; Robbs, P.G.; Leite, S.G.F. (1999). Growthand chemical composition of *Spirulina maxima* and *Spirulina platensis* biomass at different temperatures. *Aquac. Int.*, 7,261-275.

[40] Pelizer, L.H. (2002). Spirulina platensis growth estimation by pH determination at different cultivations conditions. EJB Electron. J. Biotechnol., 5(3),251-257.

[41] Xue, C.; Hu, Y.; Saito, H.; Zhang, Z.; Li, Z.; Cai, Y.; Ou, C.; Lin, H.; Imbs, A.B. (2002). Molecular species composition of glycolipids from *Sprirulina platensis*. *Food Chem.*, 77,9-13.

[42] Zeng, M.T.; Vonshak, A. (1998). Adaptation of *Spirulina platensis* tosalinity-stress.
i. *Comp. Biochem. Physiol. Part A Mol. Integr. Physiol.*, 120, 113-118.

[43] Estrada, J.E.P.; Bescós, P.B.; Fresno, A.M.V. (2001). Antioxidant activity ofdifferent fractions of *Spirulina platensis* protean extract. *Il Farmaco.*, 56, 497-500.

[44] Babadzhanov, A.S.; Abdusamatova, N.; Yusupova, F.M.; Faizullaeva, N.; Mezhlumyan, L.G.; Malikova, M.K. (2004). Chemical composition of *Spirulina platensis* cultivated in Uzbekistan. *Chem. Nat. Compd.*, 40(3),276-279.

[45] Cozza, K.L.; Costa, J.A.V. (2000). Lipídiosem*Spirulina*. *Vetor Rev. Cienc. Exatas Eng.*, 10,69-80.

[46] Richmond, A. (1990). *Handbook of microalgalmass culture*. CRC Press,Boston.

[47] Richmond, A. (2004). *Handbook of microalgal culture -biotechnology and applied phycology*. Blackwell Science,Oxford.

[48] Vonshak, A. (1996). *Spirulina platensis (Arthrospira) - physiology, cell biology and biotechnology*. Taylor & Francis,Londre.

Optimization of Florisynthesis of Silver Nano Particles Using Cordia Sebastena Linn.

Arulmathi Ramalingam and N.K. Udayaprakash*

Department of Biotechnology, School of Life Sciences,
Vels Institute of Science, Technology and Advanced Studies, Chennai, India.

nkudayaprakash@gmail.com

Abstract. Synthesis of Silver Nanoparticles [AgNPs] using biological method is widely studied. This mostly involves the extracts of leaves of different plants which resulted in nano/microparticle formation. This was mostly due to the presence of complex phytochemicals. However flower of the plants are found to possess minimal amount of secondary metabolites as well as carotenoid pigments which may serve as a reducing agent. Thus, in this study synthesis of Silver nanoparticles using floral extract of *Cordia sebastena* and optimization in their synthesis using different pH [5, 7 and 9], concentration of silver nitrate [1mM, 2Mm and 3mM] and floral extract [0.5:10, 1:10 and 1.5:10] was conducted. The characterization of AgNPs was carried using UV-Visible Spectrophotometer, Fourier Transform Infra Red [FTIR] Spectrophotometer and Dynamic Light Scattering [DLS]. Further, the floral extract was screened for the presence of different phytoconstituents. The study showed varied Plasmon resonance ranging from 430 to 470nm. Fourier transform infrared spectrum reveals the presence of phytochemicals which acted as reducing and stabilizing agent. The presence of Alkaloids, flavonoids, Phenols, tannins, cardiac glycosides, terpenoids and pholobatannins were recorded in the extract of the flower of *Cordia sebastena.* The study confirmed the synthesis of AgNPs with the average particle size of 85.2nm. The optimized condition revealed that the flower extract added with 1: 10 concentration to 2mM silver nitrate at pH9 gives smaller AgNPs.

Key Words: Silver nanoparticles, Biosynthesis, Optimization, *Cordia sebastena,* Florisynthesis, pH

1. Introduction

Nanotechnology is an interdisciplinary study of science that merges biology, physics and chemistry. When compared with microparticles, nanoparticles hold more surface area that improve its functional competence. A range of metals are exploited in the synthesis of nanoparticles which includes Au, Ag, Cu, Zn, Fe and Pd. The multifunctional potential of these nanoparticles include copious applications in various fields such as medicine, agriculture, food and textile industry. Though many metals were studied, silver nanoparticles [AgNPs] have reached a unique spotlight because of its characteristic properties, like electrical conductivity, chemical stability and surface Plasmon resonance properties and apart from that AgNPs has its extensive biological applications.

Green synthesis of nanoparticles has received an extensive interest in science and technology. Green synthesis of silver nanoparticle is sprouting into a vital branch of nanotechnology in recent decades for its huge applications which includes Anti-angiogenic Properties [1], Antimicrobials [2], Antioxidants [3], cytotoxic [4], Dye degradation [5], Anti-inflammatory, Anticancer [6] and

Larvicidal activitiy [7]. Even though there are countless ways available for the synthesis of AgNPs, flower mediated biosynthesis is striking due to the presence of unique pigments in flowers. Our previous studies [8] on synthesis of AgNPs using plant parts resulted in formation of microparticles which was due to the presence of secondary metabolites. Thus, plant part possessing lesser amount of secondary metabolites are probed and this study was conducted using a flower of *Cordia sebastena.* Flowers are inexpensive and abundantly available and function as a biofactory with diverse organic compounds possessing flavonoids and carotenoids [9]. In the present study, synthesis of AgNPs using the flower extract of *Cordia sebatena* at room temperature is optimized under various pH, concentration of flower extract and silver nitrate. Synthesized AgNPs were characterized using various spectral analyses like UV–Vis spectroscopy, FTIR and DLS.

2. Material And Methods

The flowers of *Cordia sebastena* were collected from Vels Institute of Science, Technology and Advanced Studies, Chennai, India. The flowers were examined for pathogenic infections and healthy flowers were selected. The corolla alone was excised and washed in sterile distilled water to eliminate the surface contaminants. The washed corolla was transferred to a coarse filter paper to absorb the water.

2.1. Preparation of floral Extract

The floral extract was prepared by boiling 100gms of flower in 1000 ml of distilled water for 10 mins. Then the extract is filtered through Whatmann no 1 filter paper. The clear extract is used for further experiments. The floral extract acquired was dark saffron in colour (Fig. 1).

Fig. 1. Flower of *Cordia sebastena* and its aqueous extract

2.2. Phytochemical screening

The aqueous extract of the *Cordia sebastena* flower was evaluated for the presence of phytochemicals like flavonoids, phenols, tannins, saponins, phlobatannins, steroids, cardiac glycosides, and terpenoids as described by Evans [10] and Udayaprakash et al [11].

2.3. Synthesis of AgNPs

IOP Conf. Series: Materials Science and Engineering **574** (2019) 012017 doi:10.1088/1757-899X/574/1/012017

Silver nitrate solution was prepared using millipore water (Milli Q, IQ7000, Merck), with the concentration of 1mM, 2mM and 3mM. This was adjusted to three different pH using 1N NaOH and 1N HCl to obtain silver nitrate with the pH value of 5, 7 and 9. The prepared silver nitrate with different pH and molarity was treated with the floral extract at the ratios of 0.5:10, 1:10 and 1.5:10 respectively.

2.4. Characterization of Silver Nanoparticle

The reduction of silver ions to silver nanoparticles was analysed using UV–Vis spectrophotometer with the wavelength ranging from 300–650 nm. Fourier transform infra-red spectroscopy (FT-IR) analysis of AgNPs were recorded in the range of 400–4000 cm-1. Further, the size of the particles formed was measured using Dynamic Light Scattering (DLS) Spectroscopy.

3. Results

3.1. Optimization of silver nanoparticle synthesis

To determine the ideal concentration of flower extract, concentration of silver nitrate and pH to synthesis silver nanoparticle were optimized. Synthesized silver nanoparticles were confirmed due to the colour change from colourless to reddish brown after the addition of aqueous extract of *Cordia sebastena* flowers. The characteristic colour change is noticed for the samples of pH 7 and pH 9 (Fig. 2).

Fig 2: Optimization of Florisynthesis of AgNPs at different pH, concentration of silver nitrate and floral extract of *Cordia sebastena*

3.2 UV-vis Spectroscopy

Although all samples were analysed for UV-spectroscopy, only five samples showed the characteristic peak from 430 to 470nm range. The spectral responses are inversely proportional with the aggregation of the nanoparticles formed. When the aggregation of nanoparticles takes place and the diameter increases, the peak of Plasmon resonance shifts towards broader and longer wavelengths. The UV spectrum recorded in between 430nm range to 470nm range along with the suitable condition is presented in Fig. 3.

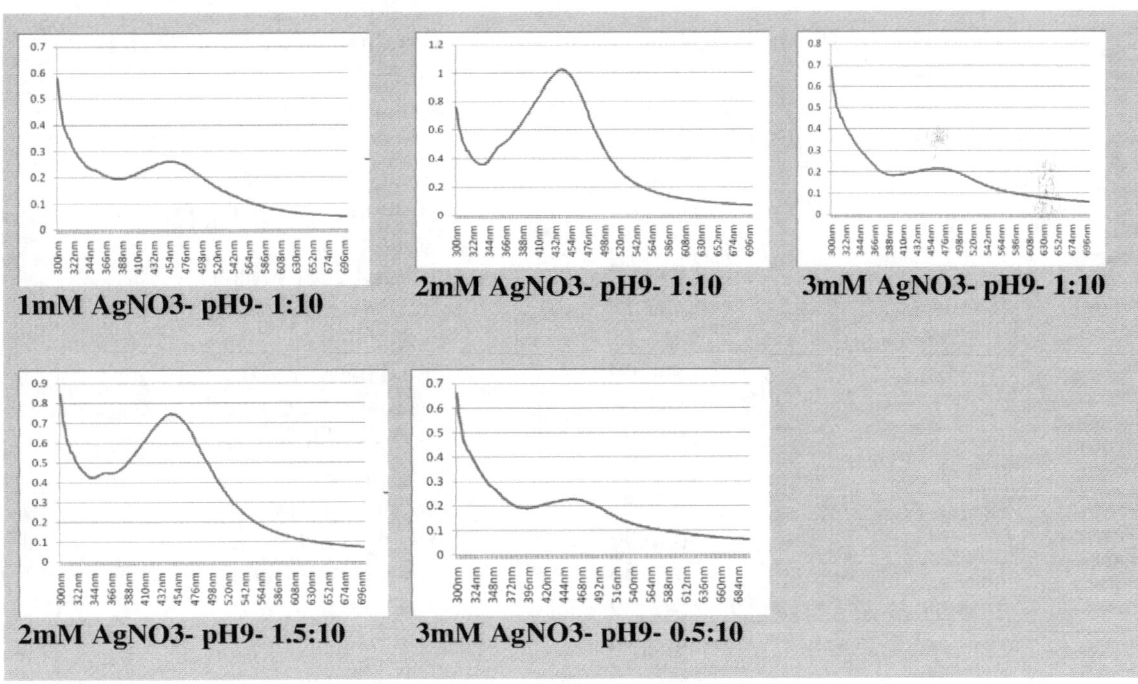

Fig. 3. UV spectrum of AgNPs synthesised using floral extract of *Cordia sebastena*

3.3 FTIR Spectral Analysis

The FTIR spectral analysis for the specific samples, i.e. 1mM AgNO3- pH9- 1:10, 2mM $AgNO_3$- pH9- 1:10, 3mM $AgNO_3$- pH9- 0.5:10, 3mM AgNO3- pH9- 1:10 and 2mM $AgNO_3$- pH9- 1.5:10 were carried. From the FTIR analysis the presence of functional groups, i.e. alcohols, alkanes, alkynes, alkenes, aldehydes, esters, aliphatic amines and carboxylic acid were confirmed. The absorption spectra and their characteristic functional groups of individual samples are presented in Table 1. The FTIR spectra recorded for different samples are presented in Fig. 4.

Table 1: FTIR Spectral Characteristic Absorption and Functional Group

Sample	Characteristic Absorption [cm-1]	Functional group
1mM AgNO3- pH9- 1:10	1080.06	C-N stretch
	1512.09	Aromatic C=C stretching
	1743.53	Aldehyde C=O stretch
	2313.46	Alkynyl C=C bending
	3186.18	Alkynyl C-H stretch
	3756.42	Amide N-H stretch
2mM $AgNO_3$- pH9- 1:10	1047.27	C-N stretch
	1171.68	Ester C-O stretch
	1745.46	Ester C=O stretch
	1424.33	Aromatic C=C stretching
	2942.21	Alkane asymmetric stretch
	3063.71	Terminal alkene =C-H stretch
	3249.43	Amine N-H stretch
	3544.92	Amide N-H stretch
3mM $AgNO_3$- pH9- 0.5:10	1013.52	C-N stretch
	1360.69	Sp^3 alkane stretch
	1422.40	Aromatic C=C bending
	2936.42	Alkane bend
	3059.38	Alkenyl C-H stretch
	3397.38	Carboxylic Acid OH stretch
3mM AgNO3- pH9- 1:10	1015.45	C-N stretch
	1360.69	Sp^3 alkane stretch
	1431.08	Aromatic C=C bending
	1625.88	C=N stretch
	1715.56	C=O stretch
	2933.53	Alkane Sp stretch
	3056.96	Alkenyl C-H stretch
	3419.56	Amine N-H stretch
2mM $AgNO_3$- pH9- 1.5:10	1011.59	C-N stretch
	1510.16	Aromatic C=C bending
	1629.74	C=N stretch
	1745.46	Aldehyde C=O stretch
	3054.07	Asymetric alkene =C-H stretch
	3438.84	Amide N-H stretch

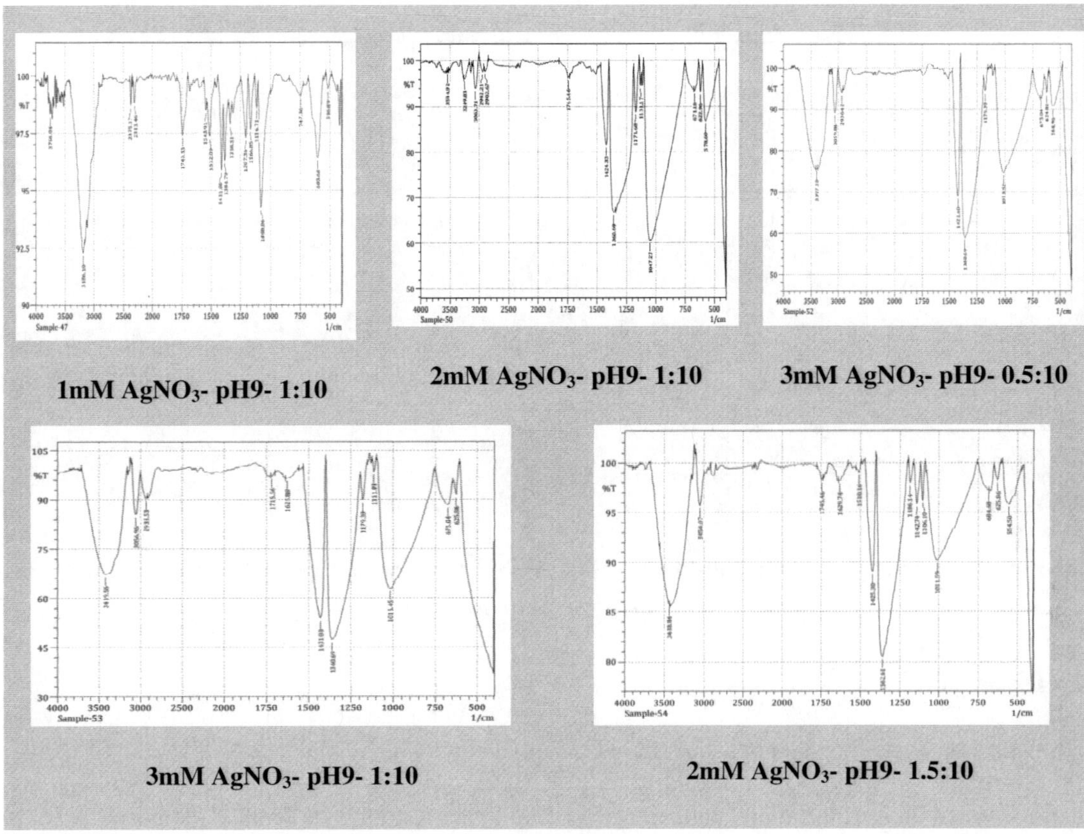

Fig 4: FTIR spectra of optimized silver nanoparticles

3.4 Dynamic Light scattering

The Dynamic light scattering studies of the specific samples, i.e. 1mM AgNO3- pH9- 1:10, 2mM $AgNO_3$- pH9- 1:10, 3mM $AgNO_3$- pH9- 0.5:10, 3mM AgNO3- pH9- 1:10 and 2mM $AgNO_3$- pH9- 1.5:10 revealed formation of AgNPs at various size. The size ranges from 1728nm to 85.2nm. The microparticles are recorded for the sample treated with 1.5 : 10 ratio of floral extract and silver nitrate. The lowest size was recorded for the sample treated with the ratio of 1:10 concentration of floral extract and silver nitrate. All the samples studied recorded their PDI index of less than 0.5. The sample, average size recorded and PDI index are presented in Table 2. The DLS spectrum and zeta potency of the sample, 2mM $AgNO_3$- pH9- 1:10 is provided in Fig. 5.

Table 2: Paricle size Distribution and PDI of synthesized nanoparticles

S.NO	Sample	Average size	PDI
1	1mM $AgNO_3$- pH9- 1:10	169.12 nm	0.481
2	2mM $AgNO_3$- pH9- 1:10	85.2 nm	0.402
3	3mM $AgNO_3$- pH9- 0.5:10	300.3 nm	0.365

4	3mM AgNO3- pH9- 1:10	119.5 nm	0.308
5	2mM AgNO3- pH9- 1.5:10	1728.1 nm	0.432

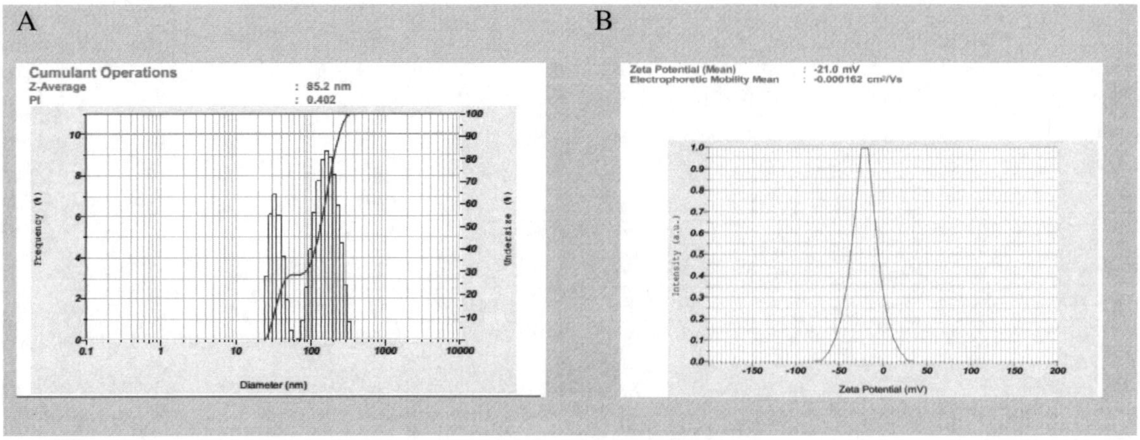

Fig 5: Particle size [a] and zeta potential [b] of AgNP synthesized using the floral extract of *Cordia sebastena*

4. Discussion

The study demonstrated the synthesis of Silver Nanoparticles (AgNPs) using the floral extract of *Cordia sebastena.* The optimization in synthesis of AgNPs was carried out using different parameters which includes, the pH, concentration of silver nitrate in synthesis of AgNPs and the concentration of floral extract in synthesis of AgNPs. The differently synthesized colloids were subjected to UV, FTIR and DLS.

UV–VIS spectroscopy is one of the best tools to examine the synthesis of the AgNPs have a property called surface Plasmon resonance which occurs is mainly due to the oscillation of the free electrons in surface of the metallic nanoparticles by any peripheral energy source[19]. Thus the physical state of AgNPs could be analysed by this optical property in the red or infrared region of the spectrum. The results of the present study confirms the fact that the alkaline pH is most suitable for the florisynthesis of AgNPs which is accordance with the reports of Dubey et al [12] and Andreescu et al [13].

FTIR gives the data of functional groups in the synthesised AgNPs for understanding their transfer of inorganic $AgNO_3$ to elemental silver by the action of the various phytoconstituents in the aqueous extract *Cordia sebastena* flowers. As known the spectra of diagnostic region excluding the fingerprint region is analysed to find the major functional groups acting on the synthesis of AgNPs. Functional group of alcohol, carboxylic acid (O-H stretching) involved in formation of AgNPs from $AgNO_3$ is in correlation with earlier report [14].

The phytochemical evaluation of the floral extract of *Cordia sebastena* resulted in detection of phenols, tannins, Alkaloids, flavonoids and terpenoids. The presence of Coumarin was reported from the flowers of *Cordia sebastena.* Among the various compounds produced by flowers, phytoconstituents such as polyphenols, flavonoids, anthocyanidins, catechins, carotenoids were participating in many bioactivities and are also known for the synthesis of silver nanoparticles [15].

The principle of DLS is to measure the fluctuations in the intensity of scattering light in the suspended particles. The Particle Dispersion Index [PDI] is used to determine the uniformity and dispersion of synthesized nanoparticle in the suspension. A broad distribution of particle is always resulted in aggregation of particle and the size and stability differs time to time. The normal Index of PDI value should always less than 0.5 [16]. The optimized condition is confirmed that the flower extract added with 1: 10 concentration to 2mM silver nitrate at pH 9 gives smaller AgNPs. The average size of the particle and the PDI value noted was 85.2nm and 0.402 respectively. Nanoparticles are defined as the particles exhibiting their size below 100 nm on any dimension [17].

5. Conclusion

The florisynthesis of AgNPs using the aqueous extract of *Cordia sebastena* was carried. Optimization on synthesis of AgNPs was performed using different parameters like pH, concentration of $AgNO_3$ and different concentration of floral extract. A total of 27 samples were analysed using UV among which only five samples have recorded the peak range between 430 – 480nm. The phytoconstituents in the *Cordia sebastena* aqueous extract were responsible for reducing and capping of AgNPs, which were confirmed by FTIR. The optimized condition is confirmed that the flower extract added with 1: 10 concentration to 2mM silver nitrate at pH 9 gives smaller AgNPs. Further studies using different flowers and other optimizing factors are recommended.

References

[1] Javad, Namvar, Ramezani, Hosseini and Mohamad. Molecules. **19**(2014): 4624-4634.
[2] Alireza E, Yahya B, Younes G and Aydin B. *Chem. Ind. Chem. Eng. Q* . **23 [1]:** (2017): 31-37.
[3] Chandrasekhar N. and. S. P Vinay. *Appl Nanosci.* (2017). DOI 10.1007/s13204-017-0624-5.
[4] Arokiyaraj S, M Valan arasu Vincent S. Prakash NU, Seong HC, Young KO, Ki and K young. International Journal of Nanomedicine. 9(2014): 379–388.
[5] Krishnaprabha M. and Manjunatha P. AIP Conference Proceedings. (2017)1832- 1836, 050016 2017; doi: 10.1063/1.4980249.
[6] Ramar M., B. Manikandan, T. Raman, K. Arunagirinathan, N.M. Prabhu, M.J. Babu, M. Perumal, S. Palanisamy, A. Munusamy. Spectrochimica Acta Part A: Molecular and Biomolecular Spectroscopy. (2014).
[7] Anjali R. International Journal of Nanotechnology and Applications.**11[1]** (2017)**:** 17-28.
[8] K. Vijayaraghavan, SPK Nalini, NU Prakash, D Madhankumar, Colloids and Surfaces B: Biointerfaces 94, (2013). 114-117
[9] Bhuvaneswari S., S. Deepa, N. Sripriya, L. Prameela, and N. K. Udaya Prakash, International Journal of Research in Pharmaceutical Sciences, 5[1] (2014). 40–45.
[10] Evans, W.C. Trease and Evans' Pharmacognosy 14th Edn. W.B. Saunders Company, London, 1996.
[11] Udaya Prakash, N.K., Bhuvaneswari, S., Balamurugan, A., Ashwin Karthick, N., Deepa, S., Hima Aishwarya, Manasveni, K.S and Sahana, International Journal of Research in Pharmaceutical Science. 4(3). (2013)476-481.
[12] Dubey, S.P., Lahtinen, M., Sillanpää, M. Colloid Surf. A,364, [1] (2010). 34–41.
[13] Andreescu, D., Eastman, C., Balantrapu, K., et al, J. Mater. Res, 22[9] (2007). 2488–2496.
[14]. Patil, M.P and G-D. Kim. Appl. Microbiol. Biotechnol. 101 (1) (2017) 79-92.

[15]. Park SY, Murphy SP, Wilkens LR, Henderson BE, Kolonel LN. Am J Epidemiol. 1732011;: 906-914.
[16] Tripathi, A., Gupta, R., Saraf, S. Int. J. Pharm. Tech. Res., 2010, 2, [3], 2116–2123.
[17] Laurent S, Forge D, Port M, Roch A, Robic C, Vander Elst L, Muller R N. *Chem. Rev.,* 108 (2008). p 2064-2110.

IOP Conf. Series: Materials Science and Engineering **574** (2019) 012018 doi:10.1088/1757-899X/574/1/012018

Comparative Analysis of Different Surface Modification Processes on AA 7075 T651

M. ABEENS*[1], R. MURUGANANDHAN[1], K. THIRUMAVALAVAN[1]

[1]Anna University, Chennai 600025, Tamil Nadu, India
*Corresponding Author address
AU FRG Institute for CAD/CAM, Department of Mechanical Engineering,
Anna University, Chennai 600025,
Tamil Nadu, India.

Corresponding Author: abeensm@gmail.com

Abstract. Modern industries turn into light weight high strength materials. Aluminum alloys are having such a great property, in which, AA 7075 T651 provides good strength than other alloys. Hence, hardness of aluminum alloys are poor there by improvement of surface hardness of aluminum alloy is necessitate. Shot peening, severe surface mechanical treatment (SSMT) and laser shock peening (LSP) are the surface property enhancement processes, which are used to improve the surface hardness of AA 7075 T651. This article deals with study the effect of shot peening, Severe surface modification process and laser shock peening on AA 7075 T 651. Among these two processes, LSP reveals better improvement in surface hardness. The Vickers micro hardness test is contacted on the specimens. 15%, 18% and 29% of harness improvements are obtain than unpeened specimen in shot peened specimen, SSMT specimen and Laser shock peened specimen respectively. The deformed layer thickness is 500 μm found in the laser shock peened specimen and 300 μm in the shot peened and SSMT treated specimens.

1. Introduction

The different surface modification processes are used to improve the surface quality of metallic materials [1-4]. On another hand, some particles like graphite, boron carbide etc., are reinforced in aluminum alloy for improving mechanical and surface properties [5, 6]. Typically, shot peening, severe surface mechanical treatment (SSMT), laser shock peening, ultrasonic shot peening (USSP) and etc. are utilized to improve the surface property of metals. During surface modification processes, permanent plastic deformation occurred on the surface of the material there by material gets hardened accordingly surface hardness of the material will be improved [7,8]. In this article has chosen to study the effects of shot peening, severe surface mechanical treatment (SSMT), and laser shock peening on AA 7075 T651.

In the shot peening process, high velocity balls are used to bombarded on the surface of the materials so the dimples are generated which reveals plastic zone and some elastic region. The recovered elastic region introduces residual stress on the surfaces [9]. Severe surface mechanical treatment (SSMT) is almost similar to the SSMT process but the high velocity is generated by the use of rotation of the drum, which surrounded by the specimen there by the impact created distance, is varying [10]. LSP process generated the Shock pressure on the metallic material surface using laser source, there by plastic deformation induced on the surface [7]. Cho et al. studied the effect of shot peening process on aluminum alloy sung Zn balls which resulted, the enhancement in hardness after shot peening [10]. Thirumavalavan K et al. reported, the SSMT process enhanced the hardness of the AA 6061 with steel balls [11]. Abeens et al. investigated the effect of LSP on AA 7075 T651 and reported significant enhancement of hardness obtained after LSP [7].

Several studies are done to enhance the hardness of AA 7075 T651 by surface modification processes; hence, there is no comparative study between surface modification processes done on AA 7075 T651. That is the aim of this research work.

2. Materials and Methods

CI steel balls (40-45 HRC) are used for shot peening and SSMT process to create significant chances on AA 7075 T651.

2.1 Materials and Methods

AA 7075 t651 material possesses high mechanical property over the others as the material is processed through thermo-mechanical treatment, where it is subjected to a heat treatment of 745K for 30 mins duration, followed by water quenching. Stretching process is done for stress relief. As a result of which an elongation of 2.5%strain elongations and the aging for 6h at 394K has been done. Due to these superior qualities, AA 7075 t651 has been used for the current experimental work [6]. The material was purchased from PCM Metals from Bangalore with 8 mm thick square plate. The specimens are cut into cuboid shape with 10 mm x 10 mm x 8 mm dimensions for Hardness. The mechanical property and chemical composition of AA 7075 T651 are given Table 1and Table 2 respectively. In shot peening process, 5 bars velocity compressive air is exerted via nozzle for bombarding balls on the surface of the material. SSMT process, the machine contains a shaft and drum. Shaft is having a specimen holder inside a drum. 100 balls are used to create impact on the surface of the material. A Q switched Nd-YAG laser with a wavelength 1064 nm, beam divergence of 0.8 mrad, beam quality factor (M2) of 2 and beam shape of top hat profile , whose diameter is 0.8 mm are used as peening parameters and 300 mJ energy is used as a laser pulse energy.

Table 1 Mechanical properties of AA 7075-T651

Property	Values
Tensile strength	540 (MPa)
Yield strength (0.2% offset)	499.87 (MPa)
Dynamic yield strength	541 (MPa)
Hugoniot Elastic Limit (HEL)	1.066 (GPa)
Elongation %	13%
Hardness (HV)	172.9 HV
Poisson's Ratio	0.33
Shear strength	330.94 (MPa)
Fatigue strength	158.58 (MPa)

Table 2 The chemical composition of AA 7075-T651

Elements	Composition
Silicon	0.03
Iron	0.14
Copper	1.3
Manganese	0.02
Magnesium	2.3
Titanium	0.05
Zinc	5.6
Chromium	0.9
Others	0.15
Aluminum	90.22

IOP Conf. Series: Materials Science and Engineering **574** (2019) 012018 doi:10.1088/1757-899X/574/1/012018

2.2 Hardness

Wilson Wolpert Vickers's micro hardness test is conducted on the AA7075 t651 at the cross sectional side from the peened surface to depth of 500 μm, in each depth five readings are taken and the average value is measured, using constant load of 1.96 N (200g) for 10 s dwell time.

3. Result and discussion

The surface of the material experiences the severe plastic deformation during surface modification processes thereby the strain hardening occurred in a peened specimens, which leads to enhancing the surface micro hardness [7]. In terms to estimate the surface hardness and the depth of hardening, the surface Vicker's micro hardness test is performed to the depth along the cross-sectional direction from the top surface of the unpeened and treated specimens. As depicted in Fig. 1 and Fig. 2 it is clear that the surface micro hardness of the LSP treated specimen is increased than other specimens

The average micro hardness of the unpeened specimen is 175 HV and little changes occur towards the depth due to its age hardening process. In the LSP treated specimens, the surface hardness value reaches 225 HV. The maximum hardness is 29 % higher than the unpeened specimen. Furthermore, the shot peened specimen reveals 202 and SSMT treated specimen exhibits 208 %. This is 11% and 8 % improvement to the LSP used specimen than shot peened and SSMT used specimen respectively.

The deformed layer thickness varies some hundred micrometers into the depth of the specimen with respect to surface modification processes. LSP treated specimen shows 500 μm deformation and other treatment treated specimens revels upto 300 μm deformation.

During the strain hardening, the material gets strengthened. This strengthening occurs owing to dislocation movements where, the dislocation lines are created by the shockwaves, which start to move across the other dislocations. Therefore the material subjected to more deformation, and the number of dislocations start from the different directions at various slip planes. Finally, the dislocation density increases due to the increasing dislocation density thereby the material hardens [12, 13]. The impact of load variation also influences the dislocation movements. As the impact effect increases eventually the dislocation density also increased due to higher deformation.

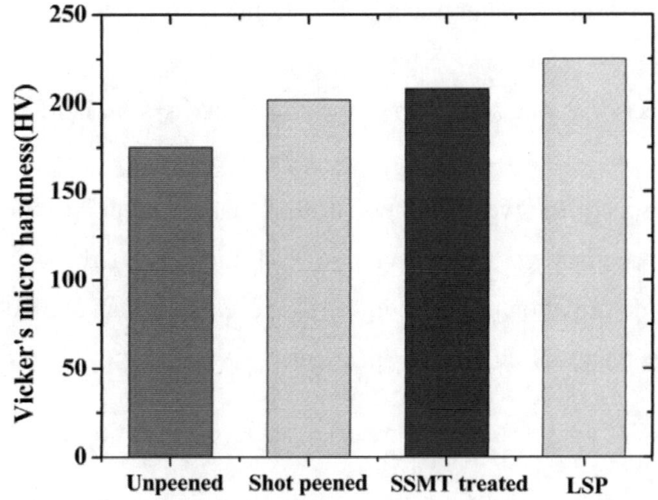

Fig. 1 Surface harness plot

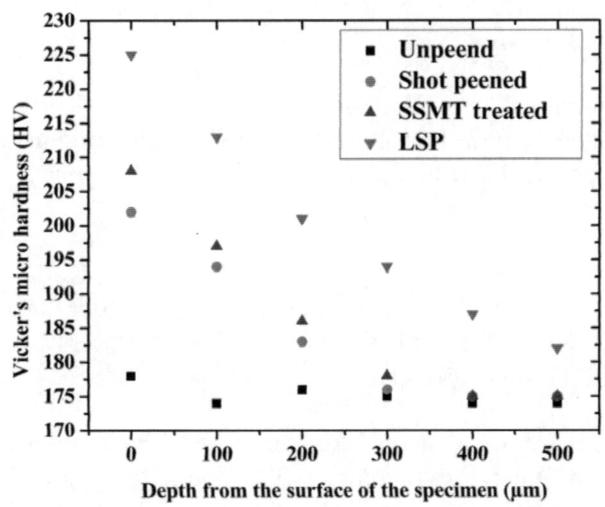

Fig. 2 Depth wise Vickers microhardness variation plot

The generated hardened layer presents some hundred micrometers depth in the LSP treated specimens. Due to, when the Shockwave propagates on the specimen the amplitude of the shock wave reduces, because it spreads to greater areas. Here the shock radius becomes higher than its initial value. When the radius reaches one-fifth higher than the initial value, the internal energy and density discontinues. This leads to the loss of the flow of the shockwave, so the hardness is decreased by the increasing depth. During the impact load, the depth of deformation increases. Due to this, the magnitude level is greater and also the strain rate is higher.

4. Conclusion

This investigation is effect of varies surface modification processes on hardness property of AA 7075 T651

- The LSP treated specimen reveals higher hardness than all other specimens, which is 8 %, 11%, 29 % enhancement compare with shot peened, SSMT and unpeened specimen.
- 500 μm depth of deformation occurred in the LSP used specimen and 300 μm deformation revealed in the Shot peened and SSMT treated specimen.

References

[1]Saravanan, M., Venkateshwaran, N.,Devaraju, A., Krishnakumari, A, and Saarvesh, J. Surface Modification Of 316l Stainless Steel ByPlasma-Assisted Low Temperature Carburizing Process, *Surface Review and Letters*, Vol. 24, No. 8 (2017): 1750116 (1-8).

[2] Saravanan, I., Elayaperumal, A., Franklin Issc, S., Vettivel, S.C. andDevaraju, A."Optimization of wear parameters and their relative effects on TiN coated surface against Ti6Al4V alloy", *Materials& Design*, Vol. 92, (2016): 23-35.

[3] Devaraju, A., Elayaperumal, A., Venugopal, S., Satish V. Kailas. and Alphonsa, J. "Investigation on the High Vacuum Tribological Characteristics of Surface Treated Nuclear Grade Stainless Steel Type AISI 316LN at 25 ºC -500 ºC", *Strojniški vestnik – Journal of Mechanical Engineering*, Vol. 57, (2011): 927-935.

[4] Devaraju, A.,Elayaperumal, A., Venugopal, S., Satish V. Kailas. and Alphonsa, J. "Microstructure and Dry Sliding Wear Resistance Evaluationof Plasma Nitride Treated Austenitic Stainless Steel Type AISI 316LN against Different Sliders", *Surface and Coatings and Technology,* Vol. 207, (2012): 406-412.

[5]Ashok Kumar, R., Devaraju, A and Arunkumar, S. "Experimental Investigation on Mechanical Behaviour and Wear Parameters of TiC and Graphite Reinforced Aluminium Hybrid Composites", *Materials Today: Proceedings, Vol. 5 (*2018): 14244-14251.

[6] Devaraju, A.,and Pazhanivel, K. "Evaluation of Microstructure, Mechanical and Wear Properties of Aluminium Reinforced with Boron Carbide Nano Composite*", Indian Journal of Science and Technology*,Vol. 9(2), (2016): 1-6.

[7] Abeens, M., Muruganandhan, R., Thirumavalavan, K., &Kalainathan, S. Surface modification of AA7075 T651 by laser shock peening to improve the wear characteristics. Materials Research Express., (2019): Vol. 6.

[8] Abeens, M., et al. "Experimental analysis of convective heat transfer on tubes using twisted tape inserts, louvered strip inserts and surface treated tube." International Journal of Ambient Energy (2018): 1-7.

[9] Gujba, Abdullahi K., and MamounMedraj, Laser peening process and its impact on materials properties in comparison with shot peening and ultrasonic impact peening, Materials, 2014, 7(12): 7925-74.

[10] Cho, KyunTaek, et al, Surface hardening of aluminum alloy by shot peening treatment with Zn based ball, Materials Science and Engineering: A, 543 (2012): 44-49.

[11] Thirumavalavan, K., L. Karunamoorthy, and K. A. Padmanabhan. "Optimization of Process Parameters Using Taguchi Technique in Severe Surface Mechanical Treatment of AA6061, International Journal of Engineering and Technology, (2014), 6(2):1026-1032.

[12] Benedetti, M., et al. "Reverse bending fatigue of shot peened 7075-T651 aluminium alloy: The role of residual stress relaxation." International Journal of Fatigue 31.8-9 (2009): 1225-1236.

[13] Lu, J. Z., et al. "Grain refinement mechanism of multiple laser shock processing impacts on AISI 304 stainless steel." ActaMaterialia 58.16 (2010): 5354-5362.

IOP Conf. Series: Materials Science and Engineering **574** (2019) 012019 doi:10.1088/1757-899X/574/1/012019

Experimental analysis of a solar cooker with a parabolic trough enhanced with PCM based thermal storage

S.Babu Sasi Kumar[1], M. Chinna Pandian[2]
[1]Phd Scholar, St Peter's Institute of Higher Education and Research, Chennai-600 054, India,
[2]Department of Aeronautical Engineering, St Peter's college of Engineering and Technology, Chennai-600 054, India,

Corresponding author Email: sbs.kumar1212@gmail.com

Abstract. Experiments are conducted on solar cooker using solar energy. Heat collected by a fluid from the Parabolic Trough Collector (PTC) with the manual Tracking Mechanism(TM) is passed through the Storage Tank (ST) .A Phase Change Materials (PCM) in the ST products absorbed the heat in the irregular sunlight and is released constantly for cooking purposes. Based on the observation of PTC and ST, a heat transfer analysis of the Solar Cooker (SC) is carried out considering 0.035, 0.045 and 0.065 kg /sec Mass Flow Rate (MFR) of the heat transfer fluid. At 1 pm, themaximum Heat Gain (HG) of PTC receiver at the Rim Angle (RA) of 82° is found to be 2095 W, 13 KW in ST with a medium of Wastage Engine Oils (WEO) .The heat output of the WEO found to be 16% more than that of the water. The efficiency of the newly designed the solar cooker is tested at the MFR 0.035, 0.045 and 0.065 kg/sec along with storage medium. The results of the MFR of 0.035 kg /sec are seen as better than 0.045 and 0.065 kg /sec for the water and WEO medium. Hence the energy extracted is greater. The WEO is seen as more stable with smaller energy losses than the water. Stearic acid (SA) is used for PCM materials.

1. Introduction

Cooking is a very essential activity worldwide. Many have started using the energy source from liquefied petroleum gas. But, in India, only 70% of population get depend on assets.SE is the only direct precious method as light energy is rehabilitated into heat energy in the Solar Cooker (SC) The SC is the most appropriate, simplest, low charge and mass convenient method of cooking food without any fuel. Employing of SC is the best for rural household in matters of time saving, quality and safety. In this paper, I like to discuss about energy analysis of PTC-receiver, ST and CV.

Michael DiGrazia et al made an alternative solar reflector sheet of a light weight, lower cost and with durability of a polymer layer with the lamination aluminum sheets of different sizes [1]. Govindaraj et al investigated the thermal performance of PTC assisted with storage system. The result showed an increase in the incident of the beam radiation causing increase an instantaneous efficiency [2]. Senthil et al did experimental work based on thermal performance of the PTC for useful energy and heat removal factor. The conclusion was that the increase in low flow rate, the concentration ratio causes increases in the thermal efficiency of PTC [3].Mohammed Hadi Ali et al experimented on the focal length of 47.02 cm based on the planned and fabricated the PTC and obtained a maximum temperature of 90˚C at an ambient temperature of 30˚C [4]. Vinayak et al. investigated work based on the PTC receiver coil. They obtained maximum temperature 160°C and its working medium water outlet temperature was increased by 50% compared to the heat losses will be 20.25% of the receiver [5]. Mohamed et al did analysis of the heat losses from PTC and have suggested that the difference between the working fluid and ambient temperature increases the temperature losses .This could be of use as higher temperature significance to reduce heat losses by using a single cover for the collector of length 10 m or less and could be more economical than a double glazing tube [6].Atul et al effect based on with and without glass cover for the PTC receiver made of fiberglass-reinforced plastic, is found that the instant efficacy of 51% and 39% with and without glass cover [7]. Jiangfeng et al investigated the performance of the PTC receiver under different parameters like a mass flow rate, atmosphere temperature, incident angle and losses. They concluded that energy loss of solar receiver increases as the inlet, with increase in wind velocity inner diameter and the convective heat loss of glass cover. [8].Recordo et al analyzed PTC receiver model, the measured techniques for the computation of heat losses and TE under varying MFR [9].

Panwar et al investigated the initial and exit energy of the solar cooker at different applications. They found SE as the best energy source for cooking [10]. Nelson et al investigated the photo voltaic system that adjusts the double axis TS which produces more energy output compared to the absence of a tracking system. In cloudy conditions, the output of solar energy with tracking system decreases [11]. Evangelos Bellosa et al investigated the initial and exit energy performance of the PTC for various storage tanks sizes of 3m^3 to 15m$^{3.}$ The storage tank of the 9m^3 obtained medium efficiency of 37.1% and 68.7% respectively in the months of the December and June [12]. Sarayooth Vaivudh et al investigated TES for a charge process. The increase in the stored energy temperature depended on Transfer fluid temperature, flow rate and initial temperature [13].Amirham Valan et al did experiment work on the well mixed recirculation water through a closed loop hot water storage system. The work showed increase in the TES temperature from 35°C to 73.84°C at the interval of 9.30 am to 4.00 pm during the period with the average solar radiation at 699 W/m^2. Observation was made every hour to evaluate the instantaneous efficiency of the collector water gain temperature, useful heat gain and values were obtained maximum at around noon [14]. Jinjia Wei et al did work on the four different capsules (sphere, cylinder, plate and tube) to observe the phase structures of PCM. The effect of the capsule diameter and the void fraction on the shell width and TES performance were studied. The spherical capsule energy output and efficiency increased during the use of four categories of the investigation [15]. Menakshi Reddy et al did experiments based on the TES using PCM.The outcome showed at almost the same output for the Stearic Acid (SA) PCM and Paraffin wax PCM but the SA was less expensive [16]. Kumarasan et al experimentally investigated the thermal property of PCM in the LHS system. From the experiment showed phase change in the melting point at beginning, peak and enthalpy [17]. Ismail et al analyzed a numerical model for the packed bed in sensible and latent heat storage systems. The models compared the objectives of the control factors like a particle size, shape, the component material, the flow rate variations, the working fluid temperature, and the wall temperature losses [18]. Sharma et al experimented on the commercial-grade SA, acetiamide and paraffin wax PCM. During the charging process the melting temperature and specific heat of PCM were observed. SA was seen with constant phase change substances than the erythritol–PCM during the charging process has helped conclusion from solar cooking model of ETSC with PCM system. The PCM unit stored energy in daylight and cooking was done at night .The procedure was continuously implemented into the system of the solar cooking and energy obtained at night was higher than daytime cooking [19].Kamal et al tested the phase change properties of the TES. The simple method has been briefly explained the energy factors such as radial temperature flow, phase change property and heat absorption are discussed at different flow rates [20]. Mawire et al did investigation of a simulation model of oil /pebble-bed TES which stored energy utilized for indirect solar cooking. The outcome proved a variable flow rate as suitable for cooking purposes as it absorbed a large quantity of heat, producing lower heat loss and could withstand a high temperature for more than an hour [21]. Mawire et al., (2010) simulated the analysis of the initial and exit energy of the solar cooker carried out using two variable flow methods. The result demonstrates the continuous maintenance of the variable flow rate at a higher temperature compared to a constant flow rate [22].

2. Experimental investigation

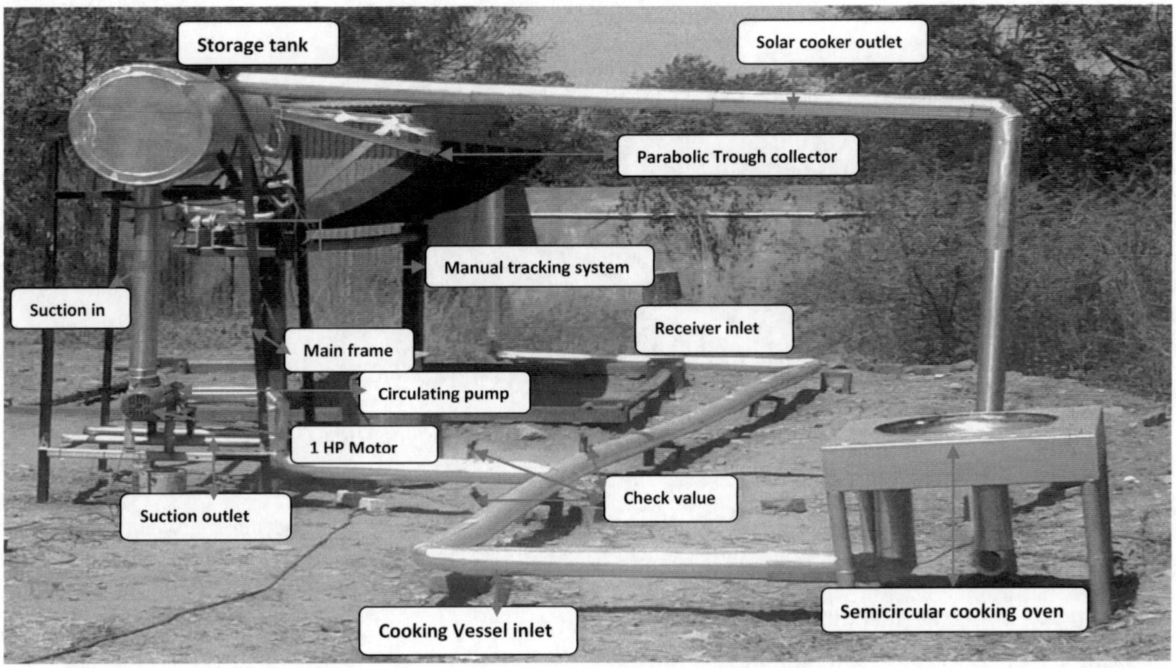

Fig 2.1 photo view of experimental setup

A photographic view of the experimental setup for SC from the PTC assisted PCM heat storage is shown in figure 2.1 .The setup consisted of two paths. Initially the transfer fluids (wastage engine oil & water) in the TES were at room temperature. This TF was pumped into the PTC receiver through the check valve for absorbing heat from the receiver and its energy was stored in the ST .The other path considered was of the stored heat transfer fluid in the ST pumped into the cooking oven for cooking purposes using the check valve. Once its utilized energy fluid was pumped back into the ST, the process was continued.

The objective of the research work is to make heat energy utilized from the fabricated PTC of the total space area 4.5 m^2 and made up of aluminium foil reflected sheet for absorbing and reflected the sun's rays focused on the receiver coil for transmitting light energy into TE by thermic fluid. The PTC receiver consisted of the stainless steel tube, vacuum glass tube, bushes and support plate. The glass tubes with one end inlet opening and the other end was sealed. Outer and interior tubes of absorber had 0.051 m and 0.043 m diameter and length (cover tube) 3 m correspondingly, the thickness of the tube was 0.4 m as shown in figure 2.2 and its specifications as shown in Table 1.

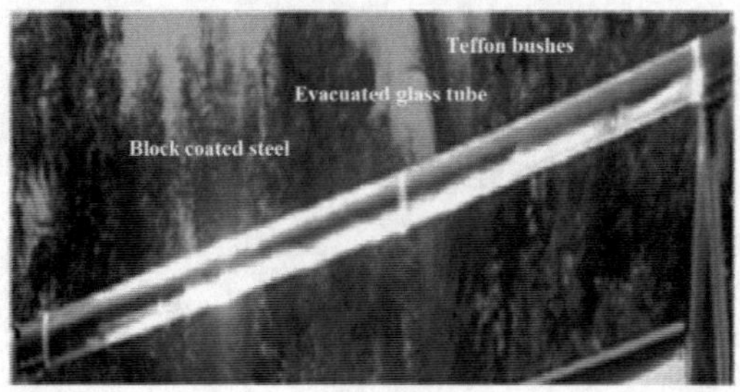

Figure 2.2 Photo view of the Receiver Tube

The inner tube-exterior surface was coated with the black paint for increasing the absorption of SR and its absorbed heat was stabilized from the air to avoid heat losses and the gap between the two tubes was sealed. The solar radiation of heat energy absorbed was converted into thermal energy by the receiver tube circulating fluid.

Table 1 PTC receiver specifications

Item	Value/Type
Collector aperture area	5.04 m^2
Aperture width	2.8 m
Receiver diameter	48 mm
Tracking mechanism type	Manually operated
Mode of tracking	Single- axis

Concentration ratio	18
Receiver length	3 m
Evacuated tube outer diameter	0.059 m
Receiver tube outer diameter	0.051 m
Evacuated tube Transmittance	0.91
ated surface absorptance	0.93
Emissivity	0.08
Pressure of Vacuum space	$5x10^{-3}$ Pa

TS were useful for continuous projection of reflected sun radiation on the focus path of the PTC. It consisted of the movement of the trackers which can have either a horizontal or an upright axis controlled by using the big worm gear supporting the rotating in the direction opposition to that of the worm as shown in the figure 2.3. The rotating worm was fixed in the coupling bush of the supporting frame. TE was used for increasing the output influence of the PTC by making the SR of the beam which focused into the receiver tube equal to zero geometrical losses. The operation of the TS system for monitoring the change in the position view point in the receiver was done every 3 minutes. The related costs, comfort design, performance and movement capability to fabricate in the manual tracking system.

IOP Conf. Series: Materials Science and Engineering **574** (2019) 012019 doi:10.1088/1757-899X/574/1/012019

Figure 2.3 Manual Operating Tracking Systems

The ST used in the experiment in the research work contained a PCM encapsulated rectangular box arranged in a parallel manner as shown in the figure 2..4. The insulated ST which contained PCM encapsulated rectangular box (size is 40 cm x 40cm x4 mm thickness) was arranged from the left to the right side of the tank connected in a parallel direction. K-Type thermocouples were fixed in between the rectangle boxes for measuring the variation in temperature of the PCM as arrangement shown in Figure 2.4. There were sixteen numbers of rectangular boxes placed in the ST for storing the heat energy and this energy can be utilized for the cooking purpose.

The encapsulated PCM contains the seal of stearic acid in the solid state. Each container has 75% SA PCM and 25% free of charge as a result of the transfer the PCM expansion in the charging and discharging processes. The TES system is fully insulated by using glass wool insulating materials. The oil flow rate system is controlled by the control valve.

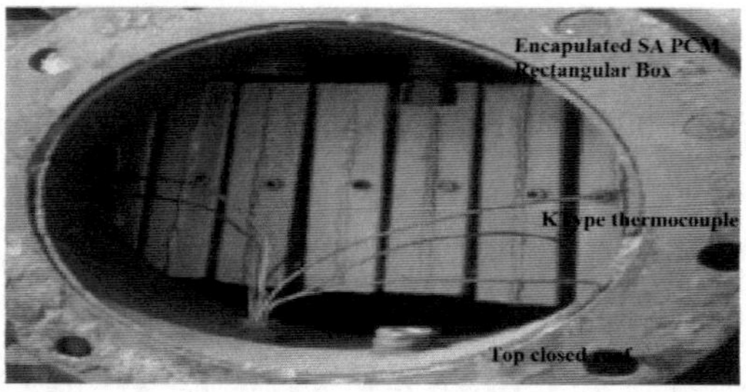

Figure2.4 Rectangular PCM boxes are arranged in the ST

The rectangular shape of the cooker consists of a semicircular shape stainless steel plate is placed on the top side, for cooking the foods as shown in the figure2.5. The bottom side of the oven is entirely covered with soldering copper coil on all sides to absorb the heat in all sides and its entire structure is insulated with the glass wool about thickness of 0.05m.

Fig 2.5 Photo view of the Cooking Vessel

Instant temperature is measured at various locations by using K-Type thermocouples. A pyrometer is a device actually tracks and measures the amount of heat that is radiated from an object. The thermal heat radiates from the object to the optical system present inside the pyrometer. The optical system makes the thermal radiation into a better focus and passes through the detector. The output of the detector will be related to the input thermal radiation. From the experiment using a MasTech (MS6610) Luxmeter.These Sensors are connected to a computer through a data logging system, which stores the outputs and records data for 30 seconds. A thermal bath, motor, pump and control valves are connected in the pipeline for charging and discharging of the PCM storage system. Here, wastage Engine Oil and Water is used as Sensible heat and PCM as a latent heat medium. Water is non-hazardous and cheapest substance. Its belongings of the specific heat, freezing point are high where its viscosity and boiling point is low.WEO becomes less viscous as it gets hotter and lubricates better and resist vaporization.SAE 10W -304-AT multi grade oils are to operate over a wide range of temperatures, so that the ambient conditions may not reach high thermal states. The efficacy of the SAE 10W -304-AT is really dependent upon the heating rates, specific heat capacities and the viscosity of the oil samples. These properties are stronger to operate at high temperatures conditions and also it provides an excellent protection against corrosive wear. The oil is suitable for all the reasons and retains in the high level of alkalinity and is used for operating cold starting and humidity condition. The experiment is conducted for different mass flow rate such as 0.035 kg/sec, 0.045 kg/sec and 0.065 kg/sec for various components of the inlet and outlet of the PTC receiver, the storage tank, the cooking vessel and PCM storage tank temperature.

3. Formula to be used:

(i)Mass flow rate of the working fluid

MFR is defined as the mass of a liquid substances passing per unit time is called as mass flow rate and is related to Volume flow rate.

$$\text{MFR} = \frac{\text{Mass in kg}}{\text{time in second}} = \frac{\text{volume in litres}}{\text{second}}$$

(ii) Thermal Efficiency (TE) of PTC:

The TE of the PTC can be simplified and defined as the ratio of useful heat (q_u) delivered per Surface Area (A) and the $I_S$as intensity of solar radiation (SR) of the surface.

Thermal efficiency $\left(\eta_{ther}\right) = \frac{q_u}{AI_S}$ --- (1)

and Heat gain $\left(q_u\right) = m\,c_p\,dT$ in Watts – (2)

Surface area $\left(A\right)$ = Width x Length in m^2

Where m is mass flow rates of working fluid in kg./sec

and I_S =Intenity of solar radiation $\left(\frac{watts}{m^2}\right)$

(iii) Useful energyoutput for the load of ST

$q_{useful} = \rho V_{st} C_p\, \Delta T$ --------- (3)

Where ρ is the density of the working fluid, $V_{st}$is the volume of storage tank, the specific heat of working fluid as C_p and ΔT is the different between the ST outlet and inlet temperature in °C.

(iv) Utilized energy output forCV at an MFR of water

The General equation of useful heat gain is

$Q = h\,A\left(T_W - T_\infty\right)$ w --- (4)

Where h is heat transfer coefficient in W/m^2K , Surface area A and T_W, T_∞ are temperature of working medium and air respectively.

$$T_F = \frac{T_W + T_\infty}{2}$$

The Reynolds number

$$R_e = \frac{V\,D}{\nu}$$

Where V for velocity of working media an D is diameter of copper tube andv iskinematic viscosity of medium .

The Nusselt Number

$Nu = c\left(Re\right)^m \left(Pr\right)^{0.333}$ ---------- (5)

Average heat transfer coefficient h $= \frac{Nu\,k}{D}$ in $W/m^{2\circ}C$

(v) Utilized energy output for CV at an MFR of WEO

The same procedure as to follows the water medium but Nu as calculated by using formula

IOP Conf. Series: Materials Science and Engineering **574** (2019) 012019 doi:10.1088/1757-899X/574/1/012019

since Re > 2300 flow is turbulent

$$\frac{L}{D} = \frac{1.3}{0.02} = 65$$

$$10 < \frac{L}{D} < 400$$

$$Nu = 0.036\left(Re\right)^{0.8}\left(Pr\right)^{0.333}\left(\frac{L}{D}\right)^{0.055} \quad ---(6)$$

4. Result and conclusion:

The experimental test was conducted in the months of February, March -2018.The designed model, geometric dimensions, geometric angles of the PTC and the Rim angle at different position of the TE and Heat gain for PTC, ST and SC and addition of PCM Storage Tank at the effect of the different MFR have been discussed. The specific heat of TF of water was seen as more than that of engine oil reflecting the variations in the efficiency and energy output. The stored energy of wastage engine oils was compared with that of water, the boiling point temperature of WEO was seen as 400°C where as the boiling point of water was 100°C, which is the fact .Hence so water got evaporated above the temperature of 100°C but the WEO stored more energy as it evaporated above 350°C.

4.1. The Performance TE of the Rim Angle

The experiment started at 9 .00 am with RA 15° and TE of 33% .When it reached 11 am, the RA was 38° at TE of 53% whereas, at 12.00 noon the, RA was 60° and TE of 69% and increased continuously depending on increase in SR. It reaches high efficiency at 1 pm, the charging efficiency of WEO at 82° RA was 78%, greater at 1 pm, after 2.00 pm it was 95° RA. There was a decrease in WEO efficiency to 75.8%. The energy was computed using equations 8 and 9. There was a decrease in the approximate constant efficiency of PTC from 1.00 pm to 2.00 pm due to the proportionate increase in the heat loss, as the energy was removed from the ST instead of being stored. Hence at 3.00 pm at 115° RA the TE decreased to 67 % as the solar radiation decreased and the heat loss increased

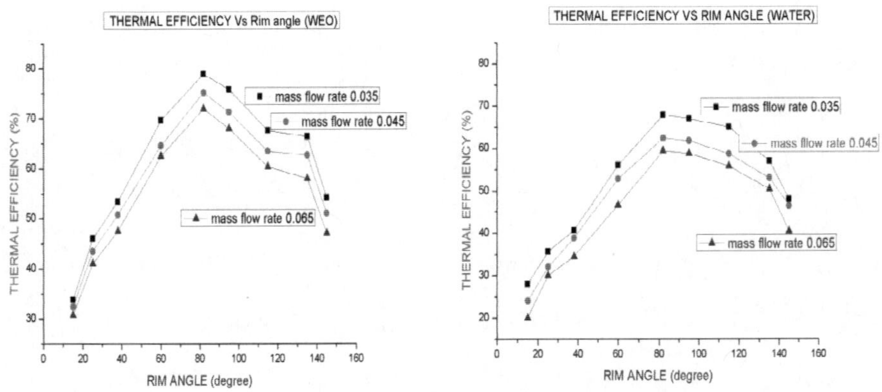

Figure 4.1 TE versus Rim Angle of the PTC

.Figure 4.1 confirms if RA as the best, TE was augmented up to 1pm and the value diminished later. At 82° RA maximum efficiency of WEO obtained was 78.7%, 75.2% and 72% for an MFR of 0.035, 0.045 and 0.065 kg /sec and 67.8%, 62.4% and 59.4% for water respectively and the performance WEO was 16%, 20% and 21% times more than that of water.

4.2 Comparing Heat gain of the PTC receiver at an MFR

Evaluate of the performance of the variation in HG of the PTC receiver with respect to time was done evaluated using the 2. At the beginning the PTC receiver temperature was increasing at a lower rate, due to the thermal inertia correlation with the system devices .

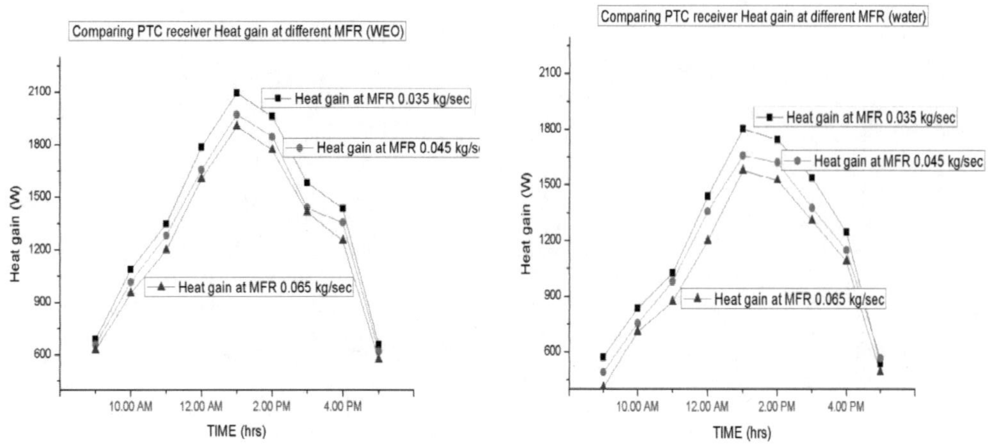

Figure 4.2 Comparing Heat gain Vs Time at an MFR the PTC receiver

Observation at 9.00 am showed the variation in ambient as 19.6°C and the intensity of the SR as 452 W/m$^{2.}$ There was an increase in radiation up to 590 W/m^2 at 1 pm. The effect of increase in the concentration ratio was high, the losses from PTC receiver showed decreases, the difference between the outlet and the inlet temperatures increased faster rate at noon due to the intensity of increase in SR and thus increasing the collection of useful energy. However, after 1.00 pm and up to 2.00 pm temperature slowed down from the peak due to a decrease in the SR. Peak efficiency was attained at 1.00 pm by which time the HG of receiver tube temperature increased. After 2.00 pm, the PTC temperature decreased due to a decrease in the system efficiency related to the increase in heat loss coefficient by convection with increase in the receiver temperature. This has been justified by Steven Boltzmann law. The result indicated a decrease in the efficiency of the receiver due to the difference in the receiver and the ambient temperatures and absorbed by employing fluid stored in the storage tank. The above discussion shows increase in the HG where by SR increasing the TF. The variation in HG increase depends on the increase in heat removal factor. After this hour, the solar radiation rate decreased and at the end of 4.00 pm it was 480 W/m^2.

At 1.00 pm with the maximum intensity of 590 W/m^2 solar radiation with 104.9°C of PTC, the Heat Gain(HG) was 1802 W and TE was 67.2% for water, for WEO when PTC receiver was 122°C, HG is 2095 W and TE was 78.7% respectively which was 16% higher than water. In the MFR 0.045 kg/sec the HG of PTC receiver is 1658 W for water, for WEO when the HG receiver is 2095 W .Thus its output is 20% more than water. After 2.00 pm its values decreased due to the absorption of heat by the working fluid passing through PTC receiver. Thus changes in SR occurred resulting in increase in the temperature of outlet working fluid based on the PTC receiver surface at the similar time due to the losses associated with an increase in environmental factors.At the MFR 0.065 kg / sec the HG is 1578 W in water medium while HG is 1904W respectively in WEO medium. The experimental value of the energy output of the WEO is 21% is shown as greater than water However, this output is less compared with MFR 0.035 kg/sec because of the flow rate and heat losses is more in the TF.

4.3 Comparing Thermal Efficiency of the PTC receiver at an MFR

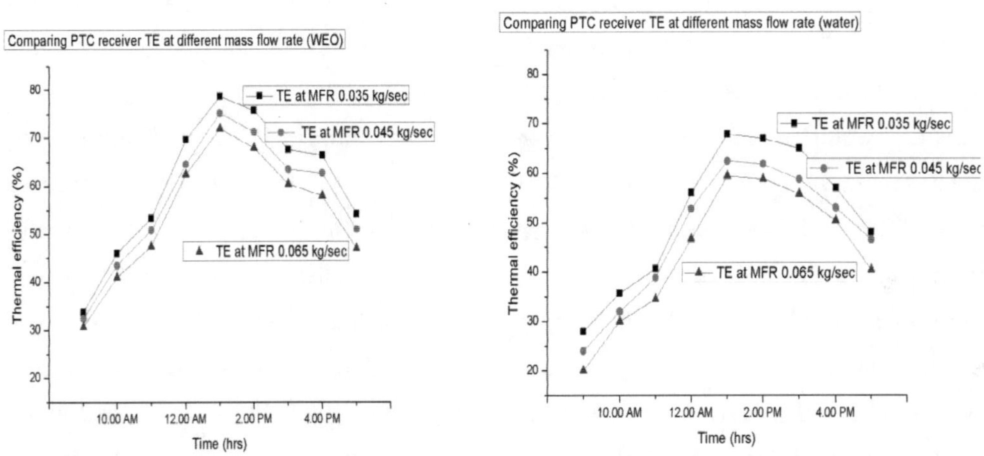

Figure 4.3Comparing TE Vs Time at an MFR of the PTC receiver

Figure 4.9 shows that, at 1.00 pm with the MFR 0.065 kg/sec PTC receiver the TE in water is 59.4% while in WEO it was 72 %. The MFR of 0.045 kg/sec PTC receiver the TE in water was 62.4% and in WEO was 75.2% but in the MFR of 0.035 kg / sec PTC receiver the TE of water was 67.2 % and WEO was 78.7 %. Thus output, shows TE of MFT 0.065 kg/sec as less compared to the usage of MFR of 0.035 and 0.045 kg/sec due to high flow rate, environmental factor and the overall heat transfer coefficient. The instantaneous efficiency of the PTC increased from 33.8 % to 46 % during the time interval between 9.00 to 10.00 am, after that there was a further increase at a faster rate until noon, and then it started decreasing. The instant efficiency of the PTC depended on both the incident SR and the HG and these two factors were more favourable during the time interval for increasing the collector efficiency.

4.4 Comparing the Performance of the ST in a transfer medium

The focus of the experiment is on the development of the energy output of the storage medium. ST which contains 16 numbers of SA PCM rectangular boxes arranged for testing under at MFR 0.035, 0.045 and 0.065 kg /sec respectively. The performance outputs of the ST during the charging and the discharging processes were determined using equations 8 and 12.At the initial stage PTC receiver, TF and ST were approximately at room temperature. After a few minutes and with an increase of convection heat from SR, storage tank temperature increased. The stored energy continued to rise up to 1 pm and decreased slowly till 3 pm. After 6 pm and without SR, heat was removed from the atmosphere and there was a rapid reduction.

TF absorbs heat from PTC receiver and exchanges the energy through PCM boxes in the charging process. The temperature of the PCM inside the rectangular box was lower than that of the melting point range, TF acted as sensible heat. In the stage of the charging process, the PCM sustained a constant temperature for a few hours. Finally the PCM got superheated and its stored energy was transmitted into a sensible heat of TF .The charging process continued till the ST temperature reached the maximum value of 110.8°C due to increased attainment of more than the atmospheric temperature difference and average heat loss per hour at noon and approximately 793 kJ/sec which increased the overall heat transfer coefficient..

When the ST reached 100°C, the overall heat loss coefficient for temperature range was 120° C to 60°C. The ST temperature decreased slowly from 1.00 pm to 3.00 pm is at regular intervals of time with its variation in the corresponding values of overall heat transfer coefficient approximately from 1.5 W/m^2K to 1.0 W/m^2K.

In the Figure 4.4 shows at 1.00 PM, at a flow rate of 0.035 kg/sec the ST output is 10237 W for water and 13655W for WEO and the energy is 1.33 times higher than the water. The comparing output of ST in water is 5.67 times and WEO is 6.5 times more than the PTC receiver at a flow rate 0.035 kg/sec. With a flow rate of 0.045 kg/sec at 1 pm of the PTC receiver HG heat of 1430 W is obtained

IOP Conf. Series: Materials Science and Engineering **574** (2019) 012019 doi:10.1088/1757-899X/574/1/012019

for water and 1953 W for WEO but in the ST it is 9987 W for water, and 12696 W for WEO medium respectively .This HG is 6 and 6.5 times higher the PTC receiver for water and WEO medium respectively.Similarly, for the useful heat output for the ST at the MFR 0.065 kg/sec of WEO and water mediaat 1.00 pm the achieved output power in ST is 9370 W, 12376 W and the energy is 5 and 6.5 times higher thanthe PTC receiver temperature of water and WEO medium respectively.After fall in SR, the storage EL for the ST is 1.08, 1.14 time each and every hour in both the water and WEO medium.

From the outcomes studied in detail and during the first 3 hours of operation, the stored energy increases due to the increase in the temperature of the flowing path of the transfer fluid in the PTC receiver and PCM ST. After 11.00 am, when the temperature of the TF reaches above 75°C approximately during the charging process for the medium of the WEO. Hence its energy is increasing continuously up to 1.00 pm with the rate of energy accumulated approximately increasesSR.At 3 pm to 5 pm, each and every an hour 0.9 times Energy Losses (EL) appears in the ST and after 6 -8 pm each and every an hour there is 1.1 times increases EL.

The rate of the energy storage is less when compared to the MFR of 0.035 and 0.045 kg/sec. The higher flow rate is engrossed when heat is draw out of the flow path and the increase heat loss is due to the environs. Therefore when the heat transfer coefficient is less PTC outlet also decreases.

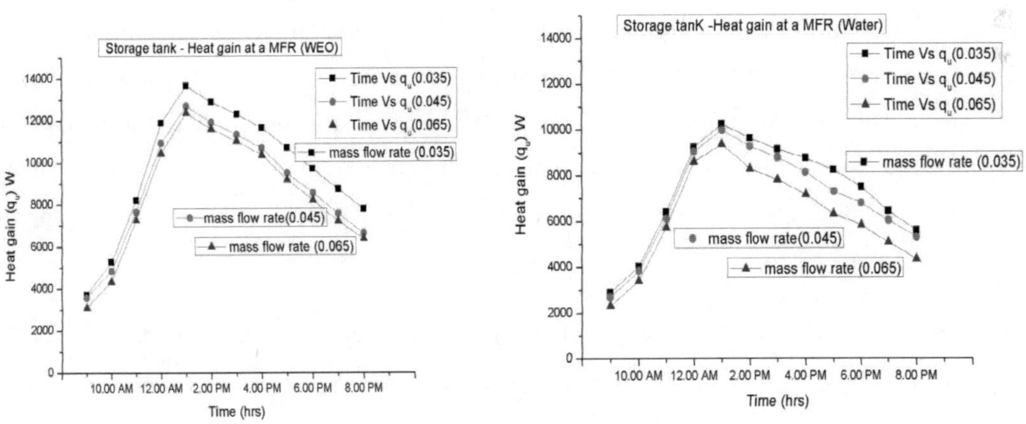

Figure 4.4 comparing the performance of the ST versus Time

Figure 4.4 is highly helpful for evaluation of the energy output of the WEO and water. At an MFR of 0.035, 0.045 and 0.065 kg/sec of the WEO strength was 1.33, 1.22 and 1.32 times higher than that the water. Further the WEO was the best for maintaining heat constantly for the charging and the disccharging processes. At 9.00 am in the MFR 0.035 kg/sec the charging energy acquired was 27% more than 0.045 and 0.065 kg/sec respectively due to the smaller amount of flow rate occupied under a larger absorption of radiation. It also reduced the environmental losses and increased the ST temperature. The experiment showed, the total charging time of the PCM material as 25% to 35% before raising the melting temperature and 20% to 25% following a rise. Therefore, the time taken for melting the PCM material of the SR was around 40% to 50% approximately.

4.5 The Presentation of the ST PCM at a different MFR

The measuring temperature allocation of the PCM rectangular box was split into six segments in the ST at x/L =0.15, 0.3, 0.45 to 0.9 meter respectively. Where L was the length of the ST in meter and was equal to 1 meter, x is the axial distance from left to right of the ST in meter and x/L is the dimensionless axial distance from the left side of the ST system. The K-type thermocouple was placed at a specific location from left to right in the ST for measuring the temperature distribution. From the beginning of the experiment, the charging temperature distribution of the ST was varied from 38° to 57°C from the right to left side and i vice-versa respectively. The heat distribution of the ST was studied before for finding the behaviour of TF. An increase in the temperature of the TF in the time gap of 9.00 to 10.00 pm through the heat gained by the PTC receiver was due to the high thermal inertia related to the various components of the experimental setup. After 10.00 am, there was a sharp increase in the temperature of the TF in the ST till 1.00 pm.This decreased at a slower rate up to 2.00 pm, after which there was no further increase in the TF heat for a period of one hour and then it started decreasing.

Let's us see from the figure 4.5 during the time interval between 10.00 am and 1.00 pm, the heat addition to the ST from PTC receiver was very high. It caused an increase in the thermal stratification between the left to the right portion of the ST. After 1.00 pm and up to 2.00 pm, the temperature distribution in the storage tank slowly decreased after 2.00 pm to 3.00 pm the charging temperature of the ST decreased quickly due to a decrease in the SR and the heat losses in the receiver due to

IOP Conf. Series: Materials Science and Engineering **574** (2019) 012019 doi:10.1088/1757-899X/574/1/012019

atmospheric condition. Later, at 3.00 pm, there was a decrease in the output energy and the average temperature of the fluid in the ST due to the density difference between the separations of the segments from left to right side of the ST. The thermal stratification between the top and bottom of the ST caused a reduction in PTC receiver temperature, SR and heat loss in RT . Figure 4.5 presents the temperature variations in ST PCM during the charge and the discharge processes for the MFR 0.035 kg/sec .SA acted as the PCM. There was a gradual increase in the PCM temperature at the beginning, at 11.00 am it reached average melting point of 69°C during which time the PCM charging was constant for half an hour and, thereafter, the charging increased sharply to attain an average TF temperature of 77°C .The charging process in the PCM segment was continuous for completing the 90% charge in all positions and the temperature in all divisions of the PCM reached an average of around 100°C, at the moment TF increased the average to around 105°C.

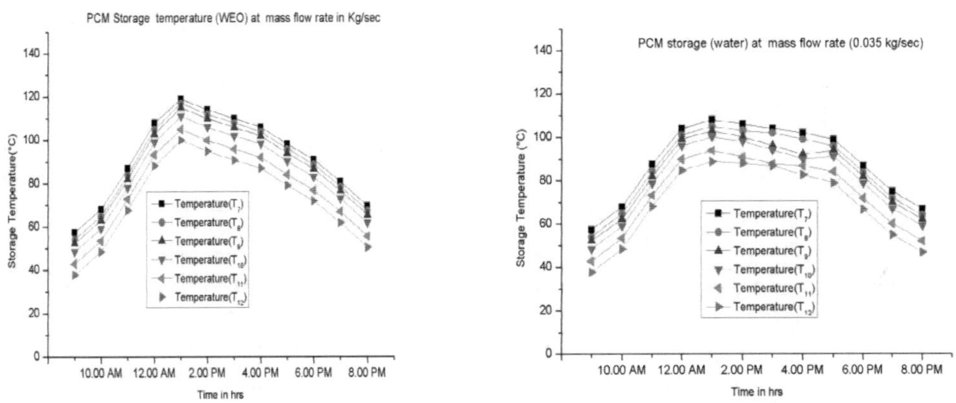

Figure 4.5 for comparing average PCM temperature Vs Time at an different MFR.

At division 2 (x/L=0.3) in the opening instant, total heat transfer of the PCM depends on the temperature of the TF alone. During sensible heating of solid PCM, the temperature of both TF and PCM got a boost at a faster rate and its difference got augmented continuously until the PCM reached the melting temperature of 69± 1°C, and the arranging process continued to reach melting temperature and the PCM heat reached was higher than TF. When the particular stage was reached, the entire heat in the TF was transferred to PCM by convection. The same procedure is followed by

IOP Conf. Series: Materials Science and Engineering **574** (2019) 012019 doi:10.1088/1757-899X/574/1/012019

another segment PCM also. After the completion of the melting process, The TF and PCM temperature continued to increase until the PCM reached the average 120°C for WEO medium.
The above study shows an increase in the storage temperature of the PCM due to the convection of TF and intensity of SR. The change in the time during the charging and the discharging processes for the water medium was not gradual but was constant in the WEO. The stored average energy of WEO was 1.15 times higher than the water medium.

4.6 Comparison of the Heat gain for SC at an MFR

The figure in 4.6 illuminated the CV utilization of the energy output at the different flow rate of the water and WEO and its output energy is computed by using the equation 4 to 6. SC outputs in the water medium were 1.02, 1.17 and 1.43 times lower than the ST at an MFR of 0.035 kg/sec, 0.045 kg/sec and 0.065 kg/sec respectively due to energy losses that happened in the atmosphere. But output in WEO was 1.06 times less than ST and this EL was continuously constant. Figure 4.6

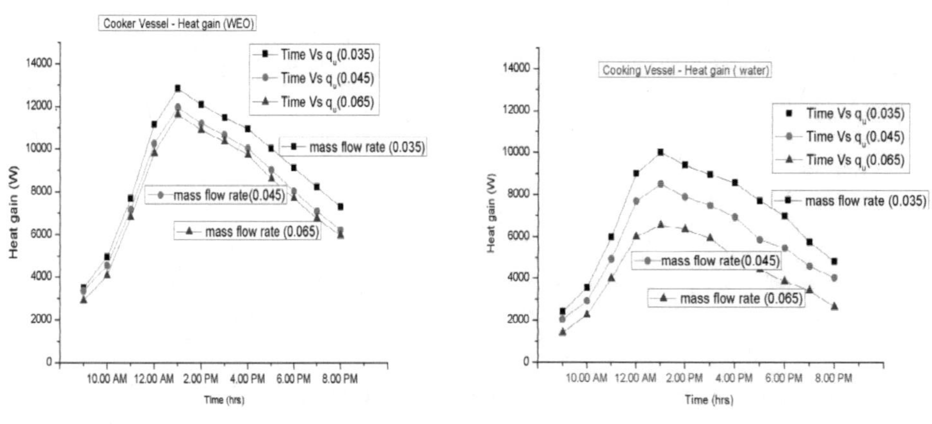

Figure 4.6 Comparison of Heat Gain Vs Time for CV at an MFR

It shows that, in the WEO medium and at a rate of 0.035 kg/sec in the SC during 6.00, 7.00 and 8.00 PM, the outputs produced were 10003, 8502 and 6537 W and its rate of 0.045 kg/sec produced output were 8031, 7036 and 6211 W respectively. Similarly for an MFR of 0.065 kg/sec the generating output were 7698 W, 6754 W and 5966 W respectively. Furthermore between 6.00 pm and 8.00 pm, the energy utilized by SC in WEO medium was 1.5 times more than the water.

5. Conclusion

This study made experimental model analysis for a solar cooker from parabolic trough collector assisted by Phase Change Materials for heat storage. Based on these aspects, Heat gain, Thermal Efficiency of the Parabolic Trough Collector-receiver, Storage Tank and Solar cooker at a mass flow rate of 0.035, 0.045 and 0.065 kg/sec were studied and outcomes are presented.

i. Increase in the solar intensity of radiation causes increase in the PTC receiver parameters like output temperature, Heat gain and Thermal efficiency are increases.
ii. The performance of the solar cooker energy output is based on various parameters like mass flow rates of transfer fluid and PCM heat storage.
iii. MFR 0.035 kg / sec is preferred to 0.045 and 0.065 kg / sec for Water and Wastage engine oil, as per the heat energy output.
iv. The Wastage engine oil is also seen as better for steady flow and energy loss than the water medium.
v. The RA of the PTC at mid position achieved a maximum HG, TE at a MFT due to an increase in the high reflectivity of radiation and outlet receiver temperature.
vi. The maximum output of the PTC receiver for WEO was reached at 1 pm at mass flow rates of 0.035 kg/sec, 0.045 kg/sec and 0.065 kg/sec respectively and the energy was 1.2 higher than that in the water medium.
vii. The energy output of Storage Tank is 6.5 times higher than PTC receiver output of the WEO, while it is 6 times higher for water medium.
viii. In the SC, energy output is less than the average of 1.1 times ST output of the WEO, water medium due to convection, conduction and radiation losses.

Abbreviations

CV	Cooking Vessel
HG	Heat Gain
MFR	Mass Flow Rates

IOP Conf. Series: Materials Science and Engineering **574** (2019) 012019 doi:10.1088/1757-899X/574/1/012019

PCM	Phase Change Materials
PTC	Parabolic Trough Collector
RA	Rim angle
RT	Receiver Tube
SA	Stearic Acid
SC	Solar Cooker
SR	Solar Radiation
SE	Stored Energy
ST	Storage Tank
TE	Thermal Efficiency
TF	Thermic Fluid
TES	Thermal Storage System
WEO	Wastage Engine oils

Nomenclature

A	Surface area, m^2
D	Receiver tube inner diameter, m
C_p	Specific heat of pressure, $Jkg^{-1}k^{-1}$
dT	difference in temperature ,°C
h	heat transfer coefficient ,$Wm^{-2}K^{-1}$
I_S	intensity of radiation Wm^{-2}
k	thermal conductivity,$Wm^{-1}{}^{\circ}K^{-1}$
L	Length in m
m	mass flow rate, kg/sec
Nu	Nusselt Number
Pr	Prandtl number
Q	Utilized heat gain in W
q_U	heat gain in w
Re	Reynolds Number
T_w	Working meiumtemperature°C
T_∞	Temperature of air °C
V	Velocity of working medium m/sec
V_{st}	Volume of Storage tank,m^3
ΔT	Temperature difference, °C
ρ	Density ,kg m^{-3}
ν	Kinematic viscosity m^2 s^{-1}
η_{ther}	Thermal efficiency ,%

References

[1] Michael DiGrazia, Gary Jorgensen 'Reflec tech mirror film: Design flexibility and durability in Reflecting solar applications,'U.S Patent, CO 80007, 2011.

[2] Govindaraj kumaresan, Rahulram Sridhar, RamalingamVelraj 'Performance studies of a solar parabolic trough collector with a thermal storage system, Energy 47(2012) 395-402.

[3] K.SenthilManikandan, G.Kumarasan, R.Velraj and S.Iniyan 'Parametric study of solar parabolic trough collector system, 'Asian journal of Applied Sciences 5(6):384-393, 2012.

[4] Mohammed Hadi Ali,' Studying & Evaluating the Performance of Solar Box Cookers (Untracked)', International Journal of Computational Engineering Research, ISSN: 2250 – 3005, Volume, 05, Issue, 08, August – 2015.

[5] VinayakSakhare, V.N.Kapatkar 'Experimental analysis of parabolic solar dish with Copper Helical coil receiver, 'International journal of innovative Research in Advanced Engineering (IIJIRAE), Volume 1, issue 8, 2014.

[6] A Mohamad, J.orfi and H.Alansary' Heat losses from parabolic trough solar collectors' international journal of energy research,'IntJ.energy Res. 2014; 38 20-28.

[7] Atul A Sagade, Satish Aher and Nilkanth N Shinde 'Performance evaluation of low-cost FRP parabolic trough reflector with mild steel receiver, 'International journal of Energy and Environmental Engineering.2013, 4:5

[8] JiangfengGuo, XiulanHuai, Zhigang Liu' Performance investigation of parabolic trough solar receiver,' Applied Thermal Engineering 95 (2016) 357–364.

[9]. Ricardo Vasquez Padilla, GokmenDemirkaya, D. Yogi GosWATERmi, Elias Stefanakos, Muhammad M. Rahman 'Heat transfer analysis of parabolic trough solar receiver,' Applied Energy 88 (2011) 5097–5110.

[10] N.L.Panwar, S.C.Kaushik, Surendra Kothari 'Solar of the art of solar cooking: An Overview,' Renewable and Sustainable Energy Reviews 16(2012)3776-3785.

[11] Nelson A. Kelly, Thomas L.Gibson 'Improved photovoltaic energy output for cloudy conditions with a solar tracking system,' Solar Energy 83 (Sep 2009), 2092–2102.

[12] EvangelosBellosa., Christos Tzivanidisa., VassilisBelessiotisb.,' Daily performance of parabolic trough solar collectors', Solar Energy 158 663–678 ,2017.

[13]SarayoothVaivudh, Wattanapong Rakwichian, Sirinuchchindaruksa, NimitSriprang 'Heat transfer

of charging and discharging experiment of high thermal energy storage system by thermal oil as Heat transfer fluid' International journal of Renewable energy,' Vol 1,No 2, July 2006.

[14]. AmirthamValanArasu and Samuel ThambuSornakumar 'Performance characteristic of the solar parabolic trough collector with hot water generation system, 'Thermal science: vol.10 (2006), No.2, pp.167-174.

[15] Jinjia Wei, YasuoKawaguchi, Satoshi Hirano and Hiromi Takeuchi (2005), "Study on a PCM heat storage system for rapid heat supply", Journal of Applied Thermal Engineering, No. 25, pp. 2903-2920

[16].R.Menakshi Reddy, N.Nallusamy, T.Hariprasad, K.Hemachandra Reddy , G.Ramachandra Reddy 'Solar energy based thermal energy storage system using phase change materials, 'International journal of Renewable energy technology,vol.3,No.1,2012.

[17] G.Kumarasan, R.Velraj, S.Iniyan' Thermal analysis of D.Mannitol for use as PCM for latent heat storage,' journal of applied science 11(16): 3044-3048, 2011.

[18] Ismail, K.A.R. and Goncalves M.M. (I999), "Thermal performance of a PCM storage unit", International Journal of Energy Conversion and Management, Vol. 40, pp. 115-138.

[19] Sharma S. D., Buddhi D. and Sawhney R.L. (1999), "Accelerated Thermal Cycle Test of Latent Heat-Storage Materials", Journal of Solar Energy, Vol. 66, No. 6, pp. 483-490

[20] Kamal A.R. Ismail and Mabruk M. Abugderahb (2000), "Performance of a thermal storage system of the vertical tube type", International Journal of Energy Conversion and Management, No. 41, pp. 1165-1190.

[21] A.Mawire, M.McPherson, R.J.van den Heet Kamp 'Simulated energy and exergy analyses of the charging of an oil-pebble bed thermal storage system for a solar cooker,' Solar energy materials and solar cells 92 (2008) 1668-1676.

[22] A.Mawire, M.McPherson, R.J.van den Heet Kamp 'Discharging simulations of a thermal energy storage system for an indirect solar cooker,' Solar energy materials and solar cells 94 (2010) 1100-1106.

IOP Conf. Series: Materials Science and Engineering **574** (2019) 012020 doi:10.1088/1757-899X/574/1/012020

An investigation on laser induced downward expanded vapour region in laser weld butt joint of AISI 316L stainless steel

A. Jayanthi [1,2], K. Venkatramanan[3], K. Suresh Kumar[*4]
[1]Research Scholar,Department of Physics, SCSVMV University, India-631 561
[2]Department of Physics, Jeppiaar Institute of Technology, India -631604
[3]Professor,Department of Physics, SCSVMV University, India-631 561
[4]P.T.Lee Chengalvaraya Naicker College of Engineering and Technology, India, 631502

*E-mail: lectsuresh25@gmail.com

Abstract. The physics of laser - AISI 316 stainless steel interactions discussed stage-by-stage including the formation of plasma plume around the keyhole during the laser welding. This work reveals that the pulsed laser heat input is a highly successful tool to join 2 mm thick 316L stainless steel welding by a single pass, that too autogenously! However, inadequacy in laser input power leads to the incomplete penetration at weld joint and poor bead formation and whereas excess heat input causes root concavity (loss of material) at the bottom. It has observed a vapour region near bottom of the keyhole while the joint fully penetrated and it slightly expanded due to excess heat input, known as downward expanded vapour region. Consequently, downward expanded vapour region results in excess melting; it leads to loss of material by spatter during welding that causes root concavity at the bottom, which will affect the joint integrity. Therefore, it is important to identify and control the expansion of vapour region by optimizing the input power of the laser. To ensure optimization of the operating parameters and the integrity of weld joints are qualified through Vickers micro hardness test and X-ray radiography test.

1. Introduction

Laser assisted material processing offers online monitoring and control of operating parameters is becoming a driver for the manufacturing industries development and sustainability [1]. Therefore, laser assisted welding became an important joining technique than the other welding techniques. Hence, laser welding will be right choice to join almost all kind of stainless steels that can promote their wide range of applications in many industries [2], because of, their excellent properties to high temperature resistance, or good corrosion resistance properties. AISI 316L stainless steels which is an extra-low carbon version of AISI 316 stainless steel, used extensively for welding where its immunity to carbide precipitation due to welding that assures optimum corrosion resistance [3]. Laser welding is an efficient process to join AISI 316L stainless steel that achieves deep penetration by forming keyhole that provides better joint strength with very small heat affected zones at rather high travel speeds. Hence, it is necessary to have a clear understanding of all the physical processes involved in keyhole welding to construct accurately placed, strong, and repeatable welds.

This article is to discuss about the formation of a keyhole in an autogenous butt joint of AISI stainless steel in the order of four states viz, solid, liquid, vapour, and plasma. The effect of increase in laser input power on AISI 316L stainless steel joints are experimented in order to optimize laser parameters for full penetration welding and discussed. The downward expanded vapour region identified on the bottom side of the welding joints and investigated.

2. **Experimental Techniques**

Sample materials considered is AISI 316L stainless steel plates having dimensions of 150 mm (length) x 50 mm (width) x 2 mm (thickness). A 500 Watts PC controlled Nd: YAG laser, wavelength of 1.064 μm, has delivered through 600-μm diameter silica core step index type optical fibre cable. The vertically focused beam on the surface operated with TEM_{00} Gaussian distribution mode and the spot diameter is 451μm at the focal length 200 mm. Nd: YAG laser source attached with highly précised CNC workbench for accurate alignments. Welding velocity fixed as 3 mm/sec on CNC machine. High purity of Argon (Ar) used as shielding gas, supplied at the flow rate 20 l/min at the upper and lower faces of the plates at an angle 45° and the nozzle placed behind the laser beam at the ambient temperature 24°C. The prepared weld sample and its operating parameters are given in figure 1 and table 1.

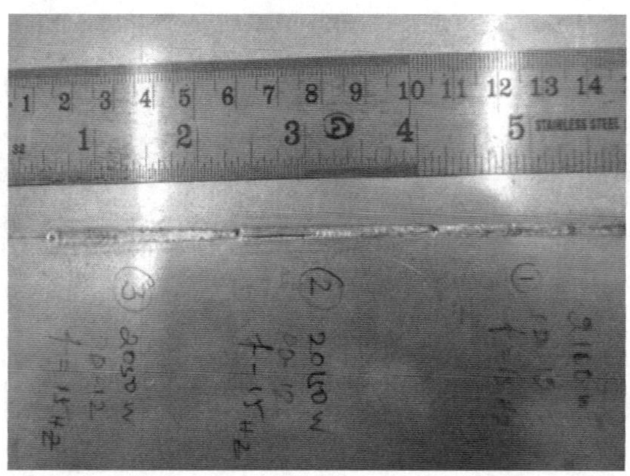

Fig.1: Photographic view of AISI 316L stainless steel butt joint welding and bead profile

Table 1: Operating parameters for pulsed laser welding

Descriptions	Values
Average Peak Power	2050- 2150 W
Spot Size	451 μm
Pulse duration	12 mS
Frequency	15 Hz
Welding Velocity	3 mm/sec
Interaction Time	48 mS

Duty Cycle	18%
Pulse shape	Recatngular pulse
pulse energy	25.8 J
Pulse over lapping %	58%

The weld joints sectioned by manual sawing are to fit in to size and shape of mounting tool. The presences of burs in the edges removed using a machine run-emery sheet. Mixture of methyl methacrylate and polyvinyl formaldehyde used as resin and hardener to mount the samples and allowed to cure up to 15-30 min. Prior to electro polishing, each specimen are grinded with 120, 240, 600 and 800 emery grits using polishing machine. 9-µm, 6-µm and 3-µm diamond slurry used to achieve fine polishing. Liquid state Mecaprex diamond compounds LD 33E for 9-µm and LDP for 6-µm are used for fine polishing respectively. Reflex 24270 used as coolant during using machine polishing. Electrolytic etching for AISI 316L Stainless Steel joints etched in a solution containing 20 ml hydrochloric acid, 1.0 g of sodium meta bisulphate and 100 ml distilled water.

Each sample are micro graphed at the range of magnification 100 µm to investigate the evolved weldpool and the keyhole regions at the cross sections of butt joint using analySIS software with PC controlled optical microscope.

3. **Results and Discussions**

3.1 Physics of laser–material interaction

In laser-material interactions, varieties of physical processes occur that are heating, melting and vaporization. According to plank's theory of radiation, atoms in a solid are not static; they are continuously vibrating about their mean positions. Laser, as an electromagnetic radiation, has an electric and a magnetic vector fields oscillating with very high frequency. If the frequency of the laser radiation does not correspond to a natural resonance frequency of the particle, then fluorescence or absorption will not occur, but a forced vibration would be initiated[4] at the surface of AISI 316L stainless steel. Hence, an average kinetic energy of these vibrations increases above the ambient temperatures roughly proportional to absolute temperature that results in increases in mean atomic distances, leading to thermal expansion against strength of the atomic bonds.However, these magnitudes of vibrations induced are very small and are incapable of vibrating an atomic nucleus. Therefore, laser irradiation assumed as photons that interacting with electrons that are either free or bound at surface. This processes of absorption photons by electrons and grow into an exited state known as "inverse bremsstrahlung effect" [5]. These excited electrons collide with lattice atoms and with other electrons and spontaneously emit its energy to all directions, since the mean free time of electrons in a conductor is 10^{-13}s. Hence, the energy transferred exited electrons to lattice atoms and transferred among the atoms via phonons that cause the lattices to vibrate beyond the activation energy (270 kJ/kg) of 316L stainless steel and small fraction atoms relocate themselves. Thus, atomic diffusion (heat flux) transmitted through space lattice as a normal diffusion-type process [6]. This cycle repeated according to the pulse repetition rate (15 Hz) with very high amount of peak power density at 1.28~1.35 MW, then, the phonon vibration increases in space lattice and atomic diffusion increases above melting point (1703 K) of 316L stainless steel [7]. Therefore, the stretched molecular bonding linkages are no longer capable of exhibiting mechanical strength, melting occurs, forms weld pool. The temperature of the heat flux beyond the weld pool decreased below melting point that forms heat affected zone. However, the range of melting and subsequent diffusivity in radial distances are limited to very few millimetres in stainless steels since, they are exhibiting high resistance to heat [8]. At this stage, most of the beam energy transferred in the direction of irradiation rather than conducts radially [9].

3.2 The process of formation of keyhole

On further heating, the temperature of the weld pool increases, the large temperature difference along the free surface of the weld pool due to the TEM_{00} mode of Gaussian distribution of laser pulses. This laser-induced thermal gradient starts to push down the molten metal under the laser beam and accelerate from a static condition to the surface of the weld pool; leads to a high temperature-dependent melt flow known as Marangoni flow. In accordance with the welding velocity (3mm/sec) and direction, the temperature gradient is very high in front of the laser spot whereas Marangoni flow is positive along the weld pool surface, which drives the molten metal to flow outwards. However, behind the laser spot where temperature gradient is lesser, the Marangoni flow is negative along the weld pool that forces the liquid metal to flow inwards. Under the hydrodynamic force and the Marangoni flow, this squeezed molten metal flows upwards along the keyhole wall, which facilitates the formation of a crater and hence the penetration depth increases, thus, a keyhole formed in the weld pool [10] as shown in figure 2. This squeezed molten metal flows upwards along the weld pool are form a humps for each pulses behind the laser spot known as weld bead as shown in figure 6.

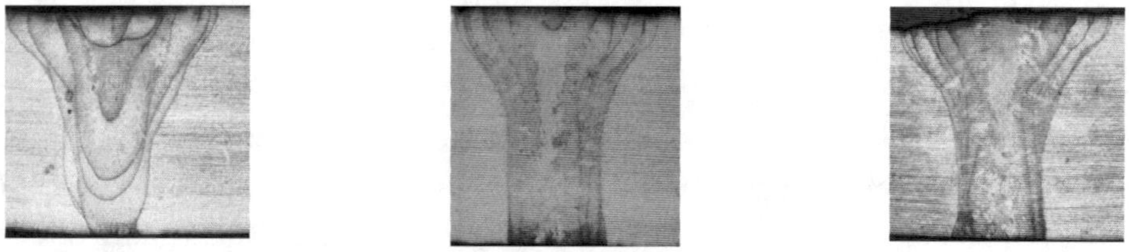

(a) (b) (c)

Fig. 2: Micrographs of cross sectional view of keyhole on 316L stainless steel at average peak power (a) 2050W, (b) 2100W and (c) 2150W.

3.3 The process of formation of plasma in keyhole

The keyhole behaves as an optical black body in the sense that radiation entering the hole is subject to multiple reflections before escaping, as a result, the absorptivity is close to 100% [11]. In such highest absorption, temperature increased above boiling point (3090 K) of 316L stainless steel and evaporation takes place by drastic increase in phonon vibration that leads to sudden increase in temperature beyond the vaporization point (5000°C) of 316L stainless steel that beaks the metallic bonds of molten material. Hence, a significant part of the absorbed energy can be directly lost by evaporating molten material and after such intensity of evaporation, recoil pressure developed. The recoil pressure brings melting point of the material closer to the boiling point by Clapeyron – Clausius principle on latent heat, hence the wall thickness of the keyhole become more thinner, which is easier to stabiles the keyhole, helps to greater penetration depth. That pushes down the liquid in the weld pool and acts as the main driving force for the keyhole formation and eventually keyhole depth increase takes which leads to even higher absorptivity due to multiple reflections. Hence, the structure of keyhole became a cavity [12], filled with mixture of vapour that surrounded by a molten region as the laser beam moves along the welding direction that forms deep penetration welds as shown in figure 3[13]. The vapour still capable of absorbing the radiation, leads to acquiring sufficient energy to induce ionization in vapour region [14]. In further irradiations, the set of the evaporated atoms with a mixture of ablated electrons and ions (electron-ion-atom) formed over the surface through the process of "inverse bremsstrahlung" phenomenon, called plasma [12] having maximum electron temperatures about 8467 K and 17,931 K[5]. Since the electron density of Argon (shielding gas) plasma is higher than the density of free electron in plasma, hence the recombination rate of electron-ion-atom lowered. So, the plasma state conserved well over the surface of the specimen for a while[15].

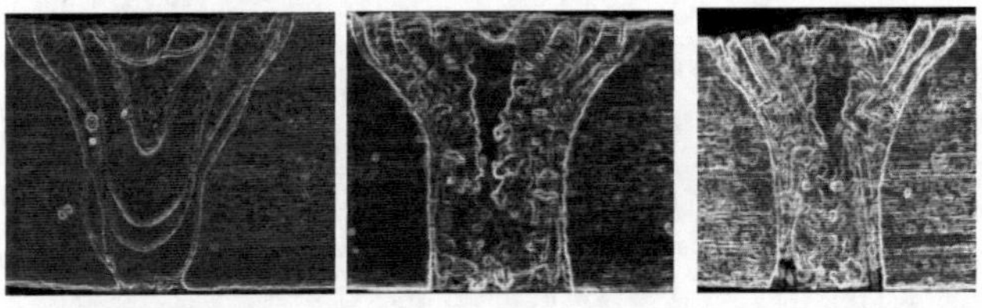

Fig.3: Micrographs of keyholes that exposing the downward expanded vapour region in 316L stainless steel by increasing from average peak power (a) 2050W, (b) 2100W and (c) 2150W.

3.4 Study of laser induced downward expanded vapour region

Laser-induced plasma plays two-fold roles during keyhole formation that the "inverse bremsstrahlung" effect absorption increases the energy coupling leading to plasma re-radiation and the absorption of reflected laser pulses from inner surfaces, called Fresnel absorption that facilitates the keyhole formation at the initial stage. In plasma, density of free electron increases and hence the current induced magnetic force increases, hence, plasma cloud start to flow that formed a vortex at the upper part of keyhole [16] as shown in figure 5. This vortex exhibit drastic change in melting point by Kerr effect and Pockel's effect; leads to a strong convectional movements in the melt pool, which has significant influence on heat transfer as discussed earlier about "Marangoni flow "[16, 17].On further increase in laser power as 2150 W,the interaction of plasma and vapour plume of keyhole generates the side flows of the melt generated beneath the focal spot by the local 'piston effect' [18]. The tilting of the keyhole front wall and the corresponding dynamic pressure of the ejected vapour plume control the degree of coupling between the vapour plume and the melt pool [5]. When the downward-flowing vapor on the front keyhole wall combined with the upward-flowing vapor, the flow field of the vapor became unstable and fluctuated rapidly. The ideal state of equilibrium of the keyhole wall was broken, and the initial wave generated on the rear keyhole wall due to the friction force between the vapour and melt flow and the fluctuating pressure of the vapor flow known as vapour-generated wave (VGW) as shown in Figure 5. From the ideal state of equilibrium, the keyhole wall appears oblique when the VGW occurred. The oblique keyhole wall indicates that the force balances are more complicated. The VGW generated in the middle of the keyhole wall moved to the bottom of the keyhole are observed as shown in micrographs (b) and (c) in figure 3. The VGW generated relative lower pressure existed at beneath of the keyhole where there is no direct laser irradiation, results in a downward-flowing vapour along the front keyhole wall. When the upward-flowing vapour of vortex met the downward-flowing vapour at the keyhole bottom generates a whirlpool and drove the suspended droplets to move in a random directions [19] as shown in the artistic image of real micrographs (b) and (c) in figure 3. Hence, the volume of the vapour region is swells known as laser induced downward expanded vapour region. It has verified with change in diameter of those keyholes obtained against the depth of the weld joints for three different laser powers as shown in table 2 and the results plotted for the comparison as shown in figure 4.

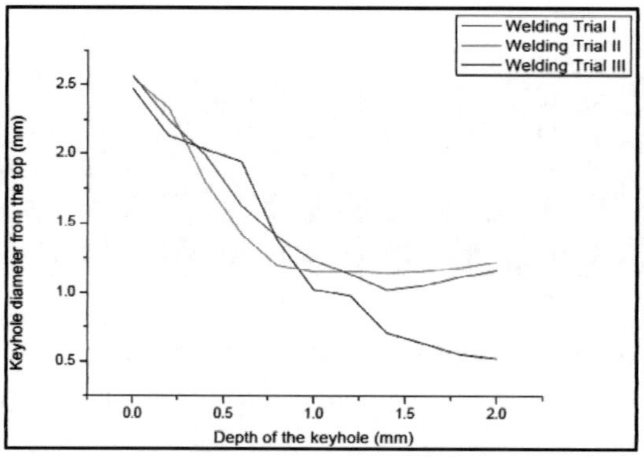

Fig. 4: Identification of downward expanded vapour of keyhole diameters in 316L stainless steel obtained for average peak power (a) 2050W, (b) 2100W and (c) 2150W.

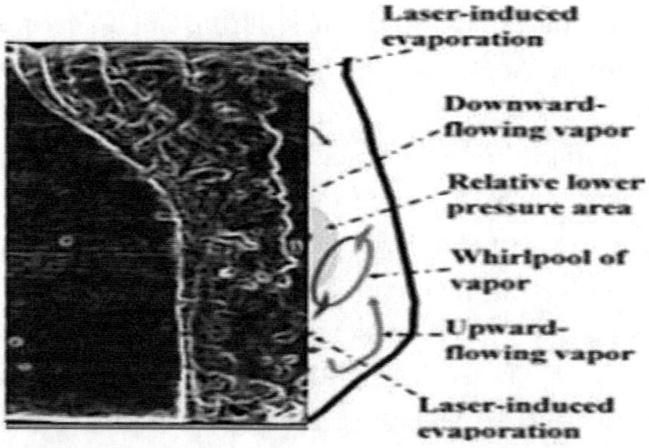

Fig.5: Comparision of generation process of the laser induced vapour region inside the keyhole wall of cross section of 316L stainless steel joint [19]

Table 2: Diameter of different keyholes of 316 L stainless steel obtained for an average peak power such as (a) 2050W, (b) 2100W and (c) 2150W

Depth of keyhole (mm)	Keyhole diameter From the top (mm)		
	Welding Trial I (2050W)	Welding Trial II (2100W)	Welding Trial III (2150W)
0.00	2.4730	2.5516	2.5665
0.20	2.1335	2.3336	2.2494
0.40	2.0321	1.7981	1.9913
0.60	1.9444	1.4228	1.6256
0.80	1.3841	1.1978	1.4062
1.00	1.0212	1.1509	1.2310
1.20	0.9784	1.1551	1.1344
1.40	0.7093	1.1440	1.0226
1.60	0.6324	1.1551	1.0516
1.80	0.5520	1.1827	1.1150
2.00	0.5244	1.2213	1.1606

Figure 4 also reveals the melt pool-HAZ boundaries of 316 L stainless steel joints and hence the swallow structure of the keyhole i.e., wider top and narrower towards bottom. The downward expanded vapour region appeared during keyhole mode welding of 316L stainless steel joints and it keep expanding when the excess heat supplied than it actually required power to achieve full penetration. However, it is interesting to point out that no signs of either vapour region or downward expanding vapour region identified from micrographs of nearly full-penetrated 316L stainless steel joint obtained for 2050W as shown in figure 3(a).

Generally, shielding gases used to protect the beam transmission by minimizing beam expansion and scattering. In the present work, high purity Argon (Ar) used as shielding gas and gas flow rate maintained at 20 l/min that is well enough to grant a suitable cost effective welding environment for stainless steels [20]. Since, Ar is a dense shielding gas flowing at the rate of 20 l/min exerts considerable amount of mechanical pressure [20] mounted on plasma that enable deep penetration and it's lessor absorption coefficient helps to protect plasma, gives best efficiency between the laser beam and specimen at the given laser power[22].

Figure 6 shows the bead morphology obtained from top and bottom of the weld joints for the different input power of the laser. Though it has reached full penetration for 2050 W, sound weld bead observed at the top and poor weld bead formed due to the insufficient melting at the bottom.

Welding trials	Top bead	Bottom bead
I (2050W)	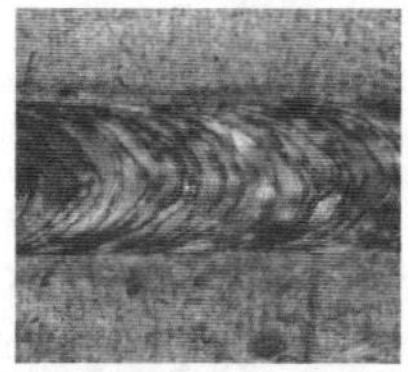	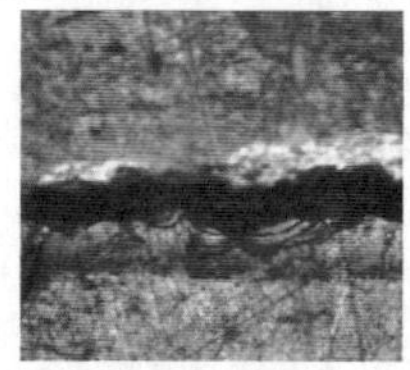
Comment Trial I:Sound weld bead found on top surface, but, incomplete Penetration at the bottom		
II (2100W)	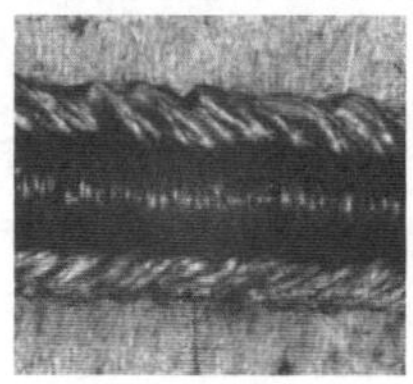	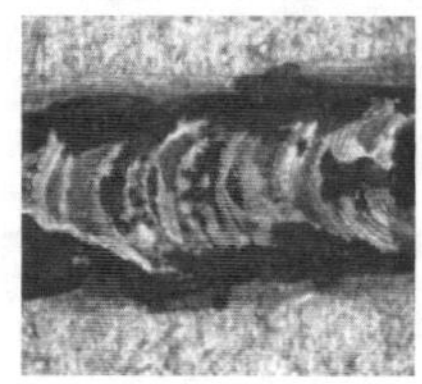
Comment Trial II: Sound weld bead found on top and bottom surface, full penetration acheived		
III (2150W)	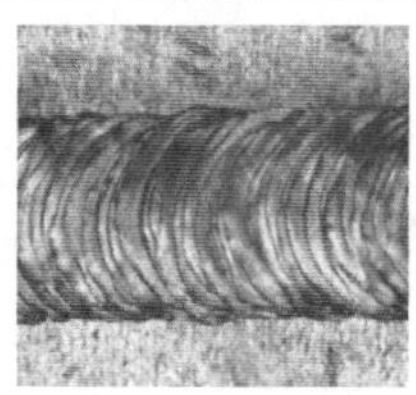	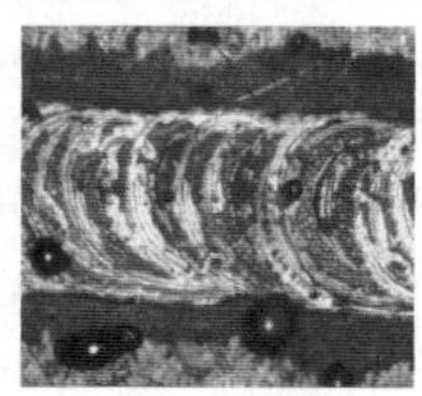
Comment Trial III: Sound weld bead found on top surface, but, spatters found at the bottom bead		

Fig.6. Weld bead morphology of 316L stainless steel joints obtained from top and bottom for different average peak power (a) 2050W, (b) 2100W and (c) 2150W.

Whereas in 2100 W, acceptable weld bead has observed at both top and bottom. At the same time in 2150 W, acceptable weld bead has observed at both top and bottom but significant spatter has observed due to the downward expanded vapour region. This leads to loss of material by spatter during welding that causes a shallow groove that may occur in the root of a butt weld known as root concavity. Root concavity caused by shrinkage of the larger weld pool due to excess input power in the through-thickness direction of the weld. In effect, the standards require that the minimum design throat thickness of the finished weldment. Therefore, it is important to identify and control the formation of downward expanded vapour region by optimizing the operating parameters, mainly by input power and welding speed.

3.5 Qualification of weld joints

3.5.1 Vickers micro hardness test

Micro hardness measured using a Vickers micro hardness tester with diamond indenter across top surface of the welded 316L stainless steel joint shown in Fig.7. In the narrow fusion line, there is an increase in hardness compared to the HAZ due to its finer microstructure around the keyhole region. It has seen that the base metal hardness is lower compared to the hardness of both HAZ and fusion zone. The hardness value decreases towards HAZ and parent metal due to the decrease in cooling rate as a function of distance from the centreline weld. The distribution of micro hardness of 316L stainless steel joint reveals the integrity of weld by exchanging molten materials from both end of the parent materials by convective flow across the fusion zone. Thus, an efficient weld joints made with different combination of 316L stainless steel plates using pulsed Nd: YAG laser source.

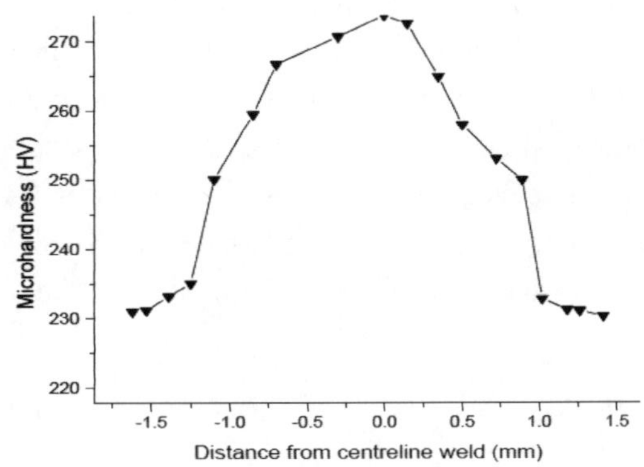

Fig.7: Micro hardness across 316L stainless steel joint welded at 2150 Watts

3.5.2 X-Ray radiography test

YXLON 225 equipment having 0.8x 0.8 mm focal spot size operated at the power supply 100 kV and 10 mA that is adequate for X–ray to test steel and its alloys up to 1 inch. X-rays exposed for 2 min. at source to film distance (SFD) 800 mm. ASTM 5 hole type and ASTM 1A 06 wire type penetrometer used as image quality indicators (IQI). The test results permanently recorded in AGFA D4 type 380x100 mm size photographic films by single wall single image technique. Radiograph of 316L stainless steel weld made by pulsed Nd: YAG laser welding as shown in Fig.8.

Fig.8: Radiograph of 316L stainless steel joint made by pulsed Nd: YAG laser welding at 2150 Watts

Figs.8 shows the radiographic test results of 316L stainless steel (SMW01), possibility of 0.5 mm dia porosity reported for the entire segment of A-B and weld spatters observed on the top surface. The slight root concavity throughout the length of the segment A-B has reported for 316L stainless steel that can be eliminate by providing suitable shroud gas or backup material at the bottom of the weld joint. Otherwise, root concavity can be eliminating by optimizing/reducing input power of the laser pulse. There is no such problem arises in 316 stainless steel weld joint made with 2100 watts laser pulse as heat input power. However, no other defects reported from the joint, hence, this results categorized as No Significant Defect (NSD) welds. Similarly, no signs of thermal induced stress reported from both the weld joints (Report no: QAD/R258/RT058/13/246/14).

4. Conclusion

AISI 316L Stainless Steel plates are butt-weld autogenously using pulsed Nd: YAG laser and hence, the systematic of formation of melt, vapour and plasma plume in the keyhole during welding are discussed. Hence, the deductions made as follows,

1. Pulsed Nd: YAG laser is a highly successful tool to join 2 mm thick 316L stainless steel welding by a single pass that too in an autogenous way!
2. Inadequacy in laser input power leads to the incomplete penetration at weld joint or poor bead formation results at the bottom.
3. It has observed that the cone shaped vapour and melt regions near the bottom of the keyhole looking slightly expanded due to a downward expanding vapour region in the case of full penetration with excess power.
4. At the same time, downward expanded vapour region at the bottom leads to loss of material by spatter that causes the root concavity along weld joint. Therefore, it is important to identify and control the expansion of vapour region by optimizing the input power of the laser.

5. Vickers micro hardness test and X-ray radiography test were carryout in order to ensure optimization of welding parameters and hence, the integrity of weld joints qualified.

6. Acknowledgements

The authors wish to thank Dr. S. Venugopal and Dr. S. Murugan, “Indira Gandhi Centre for Atomic Research, India for granting permission and providing facilities to carry out the experimental work.

7.Reference

[1] Jayanthi, A, Suresh Kumar, K, Venkatramanan K,Selection and precise application of operating parameters of Nd: YAG and other laser sources for material processing. Journal of material science and engineering. 4,188, 2015.

[2] Jayanthi A, Venkatramanan K, Suresh Kumar K, Laser beam a novel tool for welding: A review, IOSR Journal of Applied Physics. 8(6),2016.

[3] K Suresh Kumar, Numerical modeling and Simulation of a butt joint welding of AISI 316L stainless steels using a pulsed laser beam. Materials Today: Proceedings. 2, 2256-2266, 2015.

[4] Andreas Otto and Michael Schmidt (2010) LANE 2010: Towards a universal numerical simulation model for laser material processing J Phy Proc, 5:35-46.

[5] Kamel Abderrazak, Wassim Kriaa, Wacef Ben Salem, Hatem Mhiri, Georges Leaplac and Michel Autric (2009) Numerical and experimental studies of molten pool formation during an interaction of a pulse laser (Nd:YAG) with a magnesium alloy. J Opt and Laser Techno, 41: 470-480.

[6] Lawrence H. Van Vlack (1998) Elements of material science and Engineering, Addison-Wesley series in metallurgy and material engineering, Thomson press (India) Ltd.

[7] Dieter Bäuerle (2011) Laser Processing of Physics and Chemistry, Fourth Edition, springer series.

[8] Xiangzhong Jin, Lijun Li and Yi Zhang (2003) A heat transfer model for deep penetration laser welding based on an actual keyhole. Int J Heat and Mass Trans, 46:15-22.

[9] Junghwan Lim and Cho Lik Chan, (1995) Modelling deep penetration laser welding using a BEM sensitivity scheme. Engineering analysis with boundary elements, 16 (2) :93-98.

[10] Renping Wang, Yongping Lei and Yoawu Shi (2011) Numerical simulation of transient temperature field during laser keyhole welding of 304 stainless steel sheets. J Opt and Laser Techno: 43(4):870-873.

[11] Alexander F H Kalplan (2005) Model of the absorption variation during pulsed laser heating applied to welding of electronic Au/Ni-coated Cu-lead frames. J of App Sci, 241(3-4):362-370.

[12] M Schellhorn and H V Bulow (1995) Deep penetration welding using a CO laser with an unstable resonator. J Opt & Laser Techno, 27(3):191-193.

[13] John Dowden, Phiroze Kapadia and Nazmi Postacioglu (1989) An analysis of the laser-plasma interaction in laser keyhole welding. J of App Sci D, 22 (6):741.

[14] Y T Cho, W S J Na, and I Cho, (2011). Numerical analysis of hybrid plasma generated by Nd: YAG laser and gas tungsten arc. Opt & Laser Technol, 43(3):711-720.

[15] M. Vänskä and A Salminen, (2013). Laser welding of stainless steel self-steering tube-to-tube joints with oscillating mirror. Proc. of the Institution of Mechanical Engineers, Part B: J Eng Manufa 2012 226: 632-640.

[16] Jun Zhou, Hai-Lung Tsai, Pei-Chung Wang (2006) Transport phenomena and keyhole dynamics during pulsed laser welding. Transactions of the ASME, 128:680-690.

[17] W M Steen, John Dowden, Michael Davis and Phiroze Kapadia, (1988). A point and line source model of laser keyhole welding. J of App Sci D, 21(8): 1255.

[18] R Fabbro (2010) Melt pool and keyhole behaviour analysis for deep penetration laser welding. J of App Phy D, 43(44):445501.

[19] S. Li, G. Chen, M. Zhang, Y. Zhou, Y. Zhang (2014) Dynamic keyhole profile during high-power deep-penetration laser welding. J of Mat Proc Techno, 214 (3):565-570.

[20] Masao Toyoda and Masahito Mochizuki (2004) Control of mechanical properties in structural steel welds by numerical simulation of coupling among temperature, microstructure, and macro-mechanics. J Sci and Techno of Adv Mat, 5:255-266.

[21] Dominique Grevey, Pierre Sallamand, Eugen Cicala and Sorin Ingat, (2005) Gas protection optimization during Nd: YAG laser welding. J Opt & Laser techno, 37 (8): 647-651.

[22] W.J.Han, J.G.Byeon and K.S. Park (2001) Welding characteristics of the Inconel plate using a pulsed Nd: YAG laser beam. J Mat Proc Techno, 113(1-3): 234-237.

Studies on High-Temperature Corrosion resistance of low carbon Steel in Actual boiler environment

Navneetinder Singh[1], Supreet Singh[2], Manoj Kumar[3], Manpreet Kaur[4], harprabhjot singh[5], sukanta sarkar[6]
[1] Department of Automobile Engineering, Chandigarh University, Mohali, Punjab, India
[2, 3] Department of Mechanical Engineering, Chandigarh University, Mohali, Punjab, India
[4] Department of Mechanical Engineering, Baba Banda Singh Bahadur Engineering College, Fatehgarh Sahib, Punjab, India
[5]Department of Mechanical Engineering, Indian Institute of Technology, Ropar
[6] Department of Metallurgical Engineering and Material Science, IIT Bombay, Powai, Mumbai, India
Corresponding author: supreet.mech@cumail.in

Abstract : In coal based power generated plants, hot corrosion and erosion are identified as serious problems, resulting in deterioration of boiler tube. In this present study, the High-Temperature Corrosion behavior of bare SA 210 Gr. A1 steel and Friction stir processed Cr_3C_2-25(NiCr) inserted Gr. A1 steel samples were compared under cyclic condition of 1500 hours in actual both environments. The kinetics of hot corrosion was measured on the base of weight-change after each cycle. The tool rotation speed for three processed samples were recorded as 1600rpm, 2000rpm, 2400 rpm respectively with transverse speed of 30 mm/min processed by three number of passes. Microstructure evolution of base and FSPed was done by electron back scatter diffraction (EBSD). X-ray diffraction (XRD)and scanning electron microscopy (SEM) along with energy dispersive spectroscopy (EDS) were performed on the exposed sample to analyze the corrosion products. The base steel sample suffered extreme erosion-corrosion in actual boiler environment, where as processed samples exhibits better hot corrosion resistance (64%, 66%, 73%) respectively. The processed samples also showed 4 fold refined grains, as confirmed by (EBSD).

1. Introduction

Hot corrosion is the depletion or deterioration of a material subjected to elevated temperature conditions. High temperature attainment is an inputted factor in several high tech industries for example fossil fuelled power plants, refineries, petrochemical industries and coal fined thermal power plants, with an adverse effect of material degradation and oxidation resulting in reduced life of these industries[1]. In whole world, coal fired thermal plants has predominance in generation of electricity. Coal is burned inside the boiler which further boils the water, and produced steam is used to run the turbines, coupled with a generator [2].

To overcome the corrosion problem number of techniques are used, Friction Stir Processed is one of the advantageous technique which results in microstructure modification and enhanced mechanical properties [3-6].In FSP, depending on processing parameters, i.e. rotational speed,

Published under licence by IOP Publishing Ltd

transverse speed and force applied, the rotating tool induces plastic deformation on base steel. The evolution of microstructure is obtained due to material flow, high temperature in processing zone (PZ), heat affected zone (HAZ) and thermo mechanical affected zone (TMAZ).

In this study, 4 mm thick SA210 grade A1 boiler steel is Friction Stir Processed, after insertion of Cr_3C_2-25(NiCr) powder at different rotational speeds 1600rpm, 2000rpm, 2400rpm, with three numbers of passes and transverse speed of 30 mm/min. samples are named as S1-1600rpm, S2-2000rpm, and S3-2400 rpm respectively.

2. Experimental Procedure

2.1. Material

In Current study the SA210 grade A1 boiler steel was selected as the base material. The chemical composition of the steel is shown in table 1. The embedded powder was Cr_3C_2-25(NiCr) with particle size (45±15 μm). The steel plates were cut down into specified dimensions i.e. 76mm ×50mm×4mm, so that the plates can be mounted in specified designed fixture for further processing.

Table 1. Chemical Composition of Boiler tube steel, ASTM-SA210 Grade A1Steel

Elements	C	Mn	Si	S	P	Fe
Wt %	0.27	0.93	0.1	0.035	0.035	Balance

2.2. Friction Stir processing

FSP was performed on vertical milling centre VMC (make viper 950, India). The tool used during the process is pinless, made up of tungsten carbide with diameter of 10 mm throughout the experiment. Fixture along with the hold sample was bolted on the VMC bed. FSP was done on the Cr_3C_2-25(NiCr) powder embedded samples with variable rotational speed of 1600rpm, 2000rpm, 2400rpm for three different plates named as S1, S2 and S3 respectively shown in Figure 1. The feed rate of 30 mm/min and plunger depth 1mm is fixed for all the three samples. Maximum 3 passes were performed on all the samples with exact overlapping on the previous pass. After processing, the specimens were rapidly cool down with the help of dry ice.

2.3 High temperatureErosion-Corrosion tests

The specimens for the high temperature corrosion were prepared in the dimensions 10x10x4 mm from the FSPed plates. All the six sides of the samples were polished with the help of disc polishing machine from grit size 120 to 1500 grades. To get the weight gain per unit area graph all the dimension is noted and initial weight of bare and processed steel is measured.

Figure 1. Schematic illustration of friction stirs processing showing Nugget zone created by pinless tungsten carbide tool.

2.4 Erosion-Corrosion Studies in actual Coal-Fired Boiler

The processed and base samples were hung through soot blower dummy points in the low temperature zone of stage-II boiler in Guru Gobind Singh Super Thermal Power Plant (GGSSTPP), Ropar, Punjab (India) for erosion-corrosion studies under cyclic conditions. In this zone, 700 tones/hr volumetric flows of flue gas and temperature about 700±10°C were recorded. The chemical composition of the ash inside the boiler is mainly silica 55%, aluminum oxide 30% and oxides of magnesium, sulphur, calcium, sodium and potassium with % in between 0.2 to 1.5 approx. The oxidation-erosion study was conducted for 15 cycles, with 100 hour exposure for each cycle, followed by 1 hour ambient air cooling. At the end of each cycle, the hanged samples were carefully examined regarding spalling tendency, color change and then subjected to weight change measurements.

3. Results and Discussions

3.1. Microstructure

The optical microstructure of substrate SA 210 Gr. A1 steel with grain size of 25μm is consisted of light colored ferrite and dark colored pearlite grains as shown in Figure 2(c3). FSP resulted in a considerable refinement inside the Nugget Zone as confirmed by Electron Back Scattered Diffraction (EBSD) as shown in Figure 2(a-d). The FSPed steel grain size caused decrement from 25μm down to about 11.705μm, 9.447μm and 6.36μm for samples S1, S2 and S3 respectively. This means that the maximum grain size reduction is upto 4 times with the plastic deformation of FSP inside the NZ. The refinement of grains i.e. fragmentation was seen due to dynamic recrystallization as temperature was increased by friction stir processing followed by rapid cooling and severs plastic deformation and temperature during FSP and refined by the effect of severe plastic deformation. The Cr presence was confirmed, which leads to increase in hardness and -strength of FSPed steel samples. Such a grain refinement was also was also seen by authors [7-10].

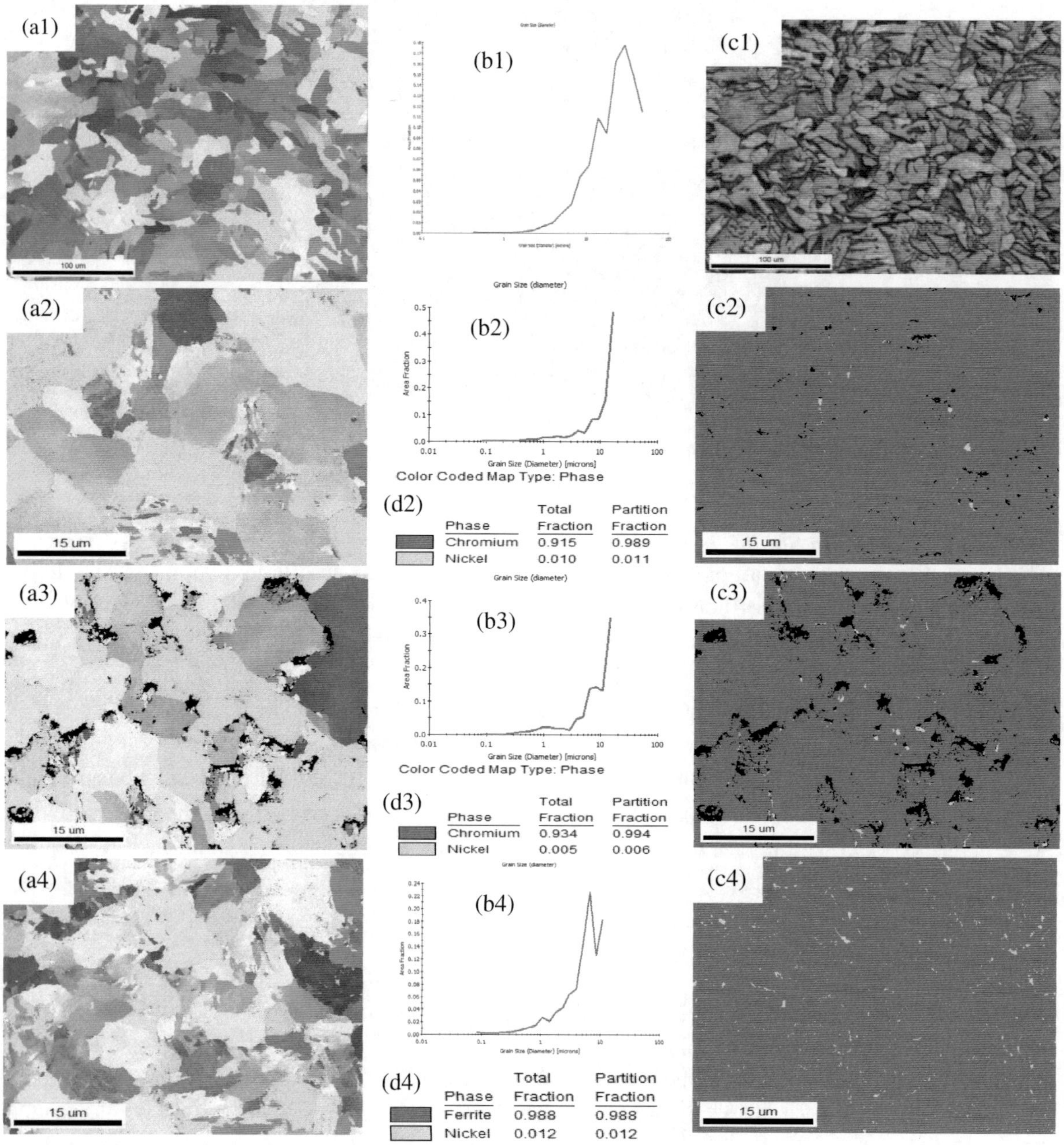

Figure 2. EBSD images showing nugget zone of (a1) base steel (a2) S1 (a3) S2 (a4) S3; (b1) base steel (b2) S1 (b3) S2 (b4) S3 graph for grain size diameter; (c2), (c3), (c4) showing % of Ni and Cr in FSPed SA 210 Gr. A1 steel; (d2), (d3), (d4) shows the chromium and nickel composition in FSPed region; (c1) shows Optical microstructure of SA 210 GrA1 boiler steel

3.2 Hot Corrosion Studies in Actual Boiler Environment

3.2.1 Visual Examination

After exposure of processed sample to actual boiler environment, due to the formation of porous oxides scale, the bare SA 210 Gr. A1 steel suffered corrosion attack. The scale appeared on the surface of the bare steel is grey in color, with settled blotches of ash (grayish brown) after 2nd cycle, At the end of 4th cycle, it is observed that on side is turned reddish brown, the predictable indication of corrosion. The corrosion was subsequently increased during cycles. By the end of 6th cycle, scale puffiness was started and uneven surface was observed, continued for 1500 hours of experimentation. An asymmetrical oxide scale was left on the bare steel sample. Cr_3C_2-25(NiCr) imbedded SA210 GrA1 steel, after exposure to boiler environment, no sign of degradation was noticed during the experimental study. A smooth and homogeneous surface was found except the patches of deposited ash. No sign of deterioration was detected on the samples even after exposure of 1500 hours.

3.2.2 Weight Change Data

The net weight loss after the exposure of samples in actual boiler environment represents the combination of weight loss and weight gain due to the spallation of the oxide scales and the formation of oxidation scale respectively. The overall weight loss by the bare steel has been found to be (-7.8824mg/cm^2). Due to formation of thick oxide scale, and continue oxidation-erosion of bare SA210 Gr. A1 steel showed higher weight loss than the Cr_3C_2-25(NiCr) imbedded steel in actual boiler environment. However, among the all Cr_3C_2-25(NiCr) imbedded, S3 sample showed the minimum weight loss (-2.12033mg/cm^2), followed by S2 (-2.62452mg/cm^2)and S1 (-2.7975 mg/cm^2) shown in Figure 3.

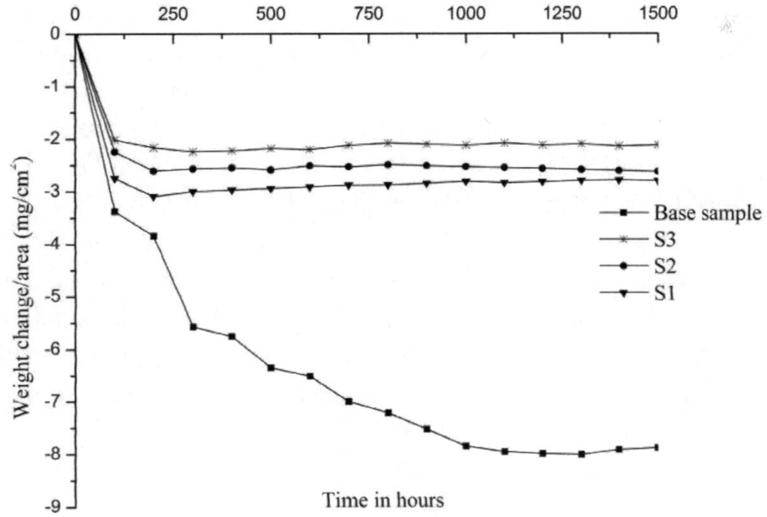

Figure 3. Weight change vs. number of cycle plots for the base sample and powered inserted FSPed samples exposed to actual boiler environment at 700 °C for 1500 hours.

3.2.3 Thickness Monitoring

After 1500 hours the 0.06433 of the thickness loss is observed in bare steel when exposed to the boiler environment .Based on this thickness data the oxidation rate will be calculated in terms of miles per year (mpy). The thickness loss data as shown in figure 4 represents the oxidation rate of SA 210 Gr A1 steel is 14.731. In the case of processed samples, thickness loss was 0.0229 mm for sample S1, sample S2- 0.0199mm, sample S3- 0.0153mm and the erosion-corrosion rate indicated by these samples were 5.224 mpy, 4.557 mpy, 3.503 mpy respectively.

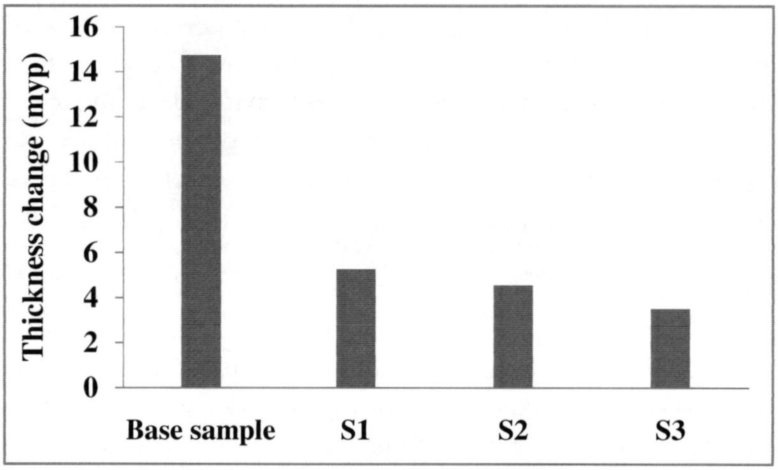

Figure 4. Bar Graph shows the thickness change in mils per year (mpy) for the Base sample and powered inserted FSPed samples exposed to actual boiler environment at 700 °C for 1500 hours.

3.2.4 X-ray Diffraction Analysis

Fig 5. shows the X-ray diffraction analysis of FSPed bare and Cr3C2-25(NiCr) imbedded SA210 Gr. A1 steel samples which were exposed on the coal-fired boiler at 700°C for about 1500hr. Fe2O3, SiO2 and Al2O3 have been identified as the strong intensity phase in bare SA210 Gr. A1 steel samples are shown in Fig. 5 (a). In the case of processed samples, for sample S1, XRD revealed the presence of Cr2O3 and Cr3C2 as the main phases with Fe2O3 and Al2O3 as weak phases (Fig. 5 (b)). For S2 sample, XRD revealed the formation of Cr2O3, Cr3C2 and Cr2C as main phases and addition of SiO2, Fe2O3 and Al2O3 phase as weak intensity phases (Fig. 5(c)). XRD analysis for the S3 revealed a prominent phase of Cr2O3, NiO, and Cr3C2 whereas SiO2 and Al2O3 phase as weak intensity phases (Fig. 5(d)). The FSPed samples oxidize along the margin of the splat during corrosion and for the processed splat FSPed samples traits to form oxidation at a rapid rate resulting in better resistance in the hot corrosion of the samples.

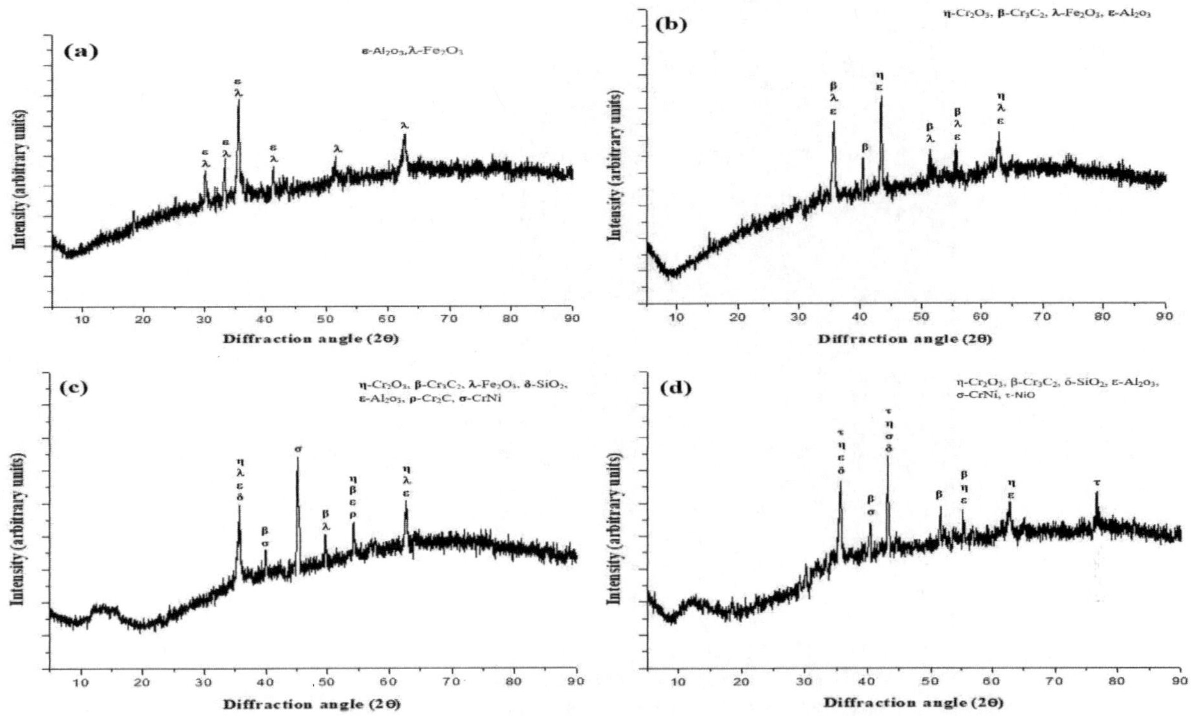

Figure 5. X-ray diffraction patterns of bare sample and powered inserted FSPed samples exposed to actual boiler environment at 700 °C for 1500 hours.

3.2.5 Surface analysis of the scale

The scanning electron microscopy (SEM) analysis of corroded boiler steel and FSPed sample S1, S2 and S3 is as shown in Figure 6. The SEM of the base sample showed (Fig 6 (a)) corroded scale with the presence of Fe_2O_3 on the surface as revealed by XRD. The presence of this harmful phases leads to formation of cracks which is formed due to penetration of more oxygen on the surface of base steel and causing high temperature corrosion. Friction Stir processed samples S1, S2 and S3 shows densely packed fiber structureon the surface. The chromium splats were seen on S3 (Fig. 6 (d)) samples as confirmed by EDS, these presences of Cr and Ni may have lead to the existence of sacrificial phases as confirmed by XRD (Fig 5 (d)). Sample S2 and S1 also showed good protection against oxidation due to strong bounded composition of Ni-Cr and capability to act as effective barriers. The SEM micrograph of the S3 sample represents inFig. 6 (d) shows us dense crack free surface. The white colored particles are observed in scale.

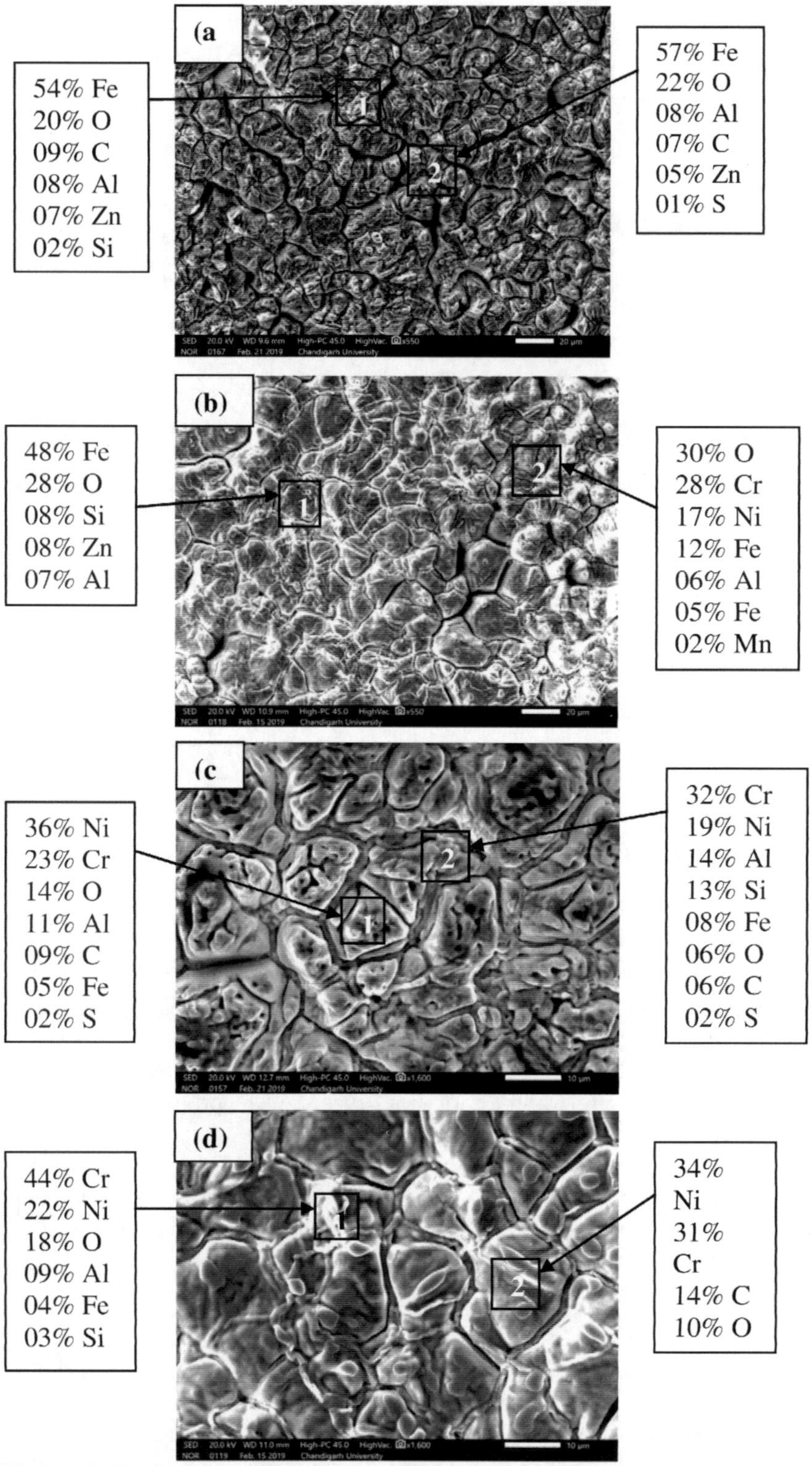

Figure 6. Surface morphology and EDS analysis for bare sample and powered inserted FSPed samples exposed to actual boiler environment at 700 °C for 1500 hours. (a) base steel (b) S1 (C) S2 (C) S3.

4. Conclusions`

- The significant refinement in the grain size is observed after FSP in the microstructure as confirmed by EBSD. The grain size in that zone decreases from 25μm down to 6.36μm this may be attributed due to plastic deformation.
- The maximum corrosion resistance was for sample S3 i.e. 73%, S3 sample showed the minimum weight gain (-2.12033 mg/cm^2), followed by S2 (-2.62452 mg/cm^2) and S1 (-2.7975 mg/cm^2). Dense surface scale,.rich in Chromium, along with significant amount of Nickel and Oxygen as revealed by the SEM microstructure, this may attribute to the probable formation of oxides and spinels of Ni and Cr.
- S3 sample revealed the presence of Cr_2O_3, NiO, and Cr_3C_2Cr and Ni may have lead to the formation of sacrificial phases and reduced corrosion rates.

References

[1] A.S. Khanna, "Introduction to high temperature oxidation and corrosion", ASM International, ISBN 0-87170-762-4, SAN: 204-7586, 2002, pp 1-322.

[2] John Stringer, "High temperature corrosion problems in coal based power plant and possible solutions", Proceedings International conference on corrosion 'CONCORN' 97, Dcember 3-6, Mumbai, India, pp 13-23.

[3] R.S. Mishra, M.W. Mahoney, S.X. McFadden, N.A. Mara, A.K. Mukherjee, ScriptaMater. 42 (2000) 163.

[4] [2] R.S. Mishra, M.W. Mahoney, Mater. Sci. Forum 357-3 (2001) 507.

[5] [3] W.M. Thomas, E.D. Nicholas, J.C. Needham, M.G. Murch, P. Templesmith, C.J.Dawes, GB Patent Application No. 9125978.8, December 1991.

[6] M. Hajian, A. Abdollah-zadeh, S.S. Rezaei-Nejad, H. Assadi, S.M.M. Hadavi, K. Chung,M.Shokouhimehr, Microstructure and mechanical properties of friction stir processed AISI316L stainless steel, Mater. Des. 67 (2015) 82-94.

[7] H.S. Grewal, H.S. Arora, H. Singh and A. Agarwal , Applied Surface Science., 268 (2013) 547-555.;

[8] H.S. Arora, H. Singh, B.K. Dhindaw, Metallurgical and Materials Transactions B: Process Metallurgy and Materials Processing Science 43 (2012) 92–108.

[9] P.Xue, B.L.Xiao, W.G.Wang, Q.Zhang, D.Wang, Q.Z.Wang and Z.Y.Ma, Mater.Sci. Eng.A 575 (2013) 30–34.

[10] P. Xue, W.D. Li, D. Wang, W.G. Wang , B.L. Xiao and Z.Y. Ma, Materials Science & Engineering., 670 (2016) 153–158.;

IOP Conf. Series: Materials Science and Engineering **574** (2019) 012022 doi:10.1088/1757-899X/574/1/012022

Reduction of NOx and HC emissions in Waste Cooking oil Biodiesel using Nanoparticles of CsTPA

Sahabdheen Abdul Basheer [1,2] Arivarasu Anitha [1]

[1] Department of Chemical Engineering, Hindustan Institute of Technology & Science, Chennai-603103, India.

[2] Department of Chemistry, Hindustan Institute of Technology & Science, Chennai-603103, India.

Corresponding Author: A. ANITHA

E-mail: arniish@yahoo.com

Mobile: +91 97912 23310

Abstract. Recyclable Nanoparticles of Cesium Tungstophosphoric acid (CsTPA) for reduced NOx emission in biodiesel using waste cooking oil of high acid value was investigated. The synthesized nano catalysts were characterized using FTIR, XRD, DLS, BET and SEM. The maximum yield of biodiesel reached 91% under optimized reaction conditions of 4% catalyst (w/w of oil), molar ratio of alcohol to oil of 10:1 for 4 hours at 65 °C. Engine performance and emission characteristics of fame acid methyl esters investigated in a four stroke diesel engine showed a decrease in the Brake Thermal efficiency and Mechanical efficiency with respect to diesel. The heteropolyacid successfully decreased the HC and NOx emissions in the biodiesel with an average reduction of 28 % and 10.3 % respectively. The nano catalyst also had excellent catalytic stability and reusability making it a promising catalyst for FAME production.

Introduction

The world has a very high energy need, of which a major percentage is fulfilled by the combustion of fossil fuels. But there are downsides to fossil fuels, as they are non-renewable and the exhaust from its combustion has a large volume of CO_2, NO_X, sulfur compounds, etc. which facilitates global warming [1, 2]. To keep up with the high energy demands and the depleting energy sources [3], prospects of tapping energy from alternate sources like nuclear, solar, hydroelectric, wind, geothermal etc. are being studied extensively. All these alternate sources are not as universal as fossil fuels as it depends on several factors (like availability of sunlight, dams, geysers, funding for nuclear plants etc.). So the need arises to explore other alternatives that are available globally. The need for an easily procurable fuel substitute led to the projection of biodiesel as a global alternative fuel with properties similar to the conventionally used fuels [4, 5].

Biodiesel is obtained by transesterifying oils and fats using alcohols like methanol and ethanol with glycerin as a by-product [2, 6, 7]. All the hurdles faced by other energy sources are overcome by biodiesel as it is carbon neutral [8], biodegradable, renewable, reduces greenhouse effect [9], produces useful by-products and can be used as such in diesel engines without any further modifications [1, 10, 11]. The catalyst used can either be a chemical or a bio catalyst or an entirely non-catalytic process for its production [3, 6].

Biodiesel has been controversial as the feedstocks used for its production are oils from plants, such as corn, soya bean, sugarcane and palm. This has led to debate over the importance of food over fuel [12]. Large quantities of land, fresh water and fertilizers are used in this process which leads to deforestation, water and soil pollution. In recent years, research is being carried out to solve this problem by using non-edible oils and oils using algae as feedstock. But all these feedstocks require more land to grow and more water resources, making the production of biodiesel to be almost economically non-viable [13].

Waste Cooking Oils (WCO) have an advantage over other oil sources, as it needs neither land for cultivation nor large quantity of water to process the feedstock and it is less costly [14]. The world produces million tons of WCO every year from various sources [15]. WCO is considered as a pollutant because vegetable oils form a very thin layer on water bodies affecting the aquatic life[5]. Therefore, use of WCO as a feedstock for biodiesel will cut down on pollution from WCO disposal and fossil fuels [11].

Homogeneous base catalysts are being widely used in transesterification reactions for synthesis of biodiesel from edible and non-edible oils [16]. Even though the alcohol requirement is low for these catalysts, the process suffers from serious drawbacks as the catalyst is intolerant towards water molecules and Free Fatty Acids (FFA) and requires high energy [17]. Water present in the raw material leads to saponification of the oil to form soap which reduces the percentage of yield. There is an increase in the volume of waste water generated using these catalysts, as the soap formed has to be removed by water washing [6]. Also, the catalyst is ineffective with oils and fats having higher Acid Value (AV) [5]. Hence, the AV of such oils and fats have to be reduced before using it for transesterification process.

Homogeneous acid catalysts like H_2SO_4 and phosphoric acid H_3PO_4 are also being used in the production of FAME[18]. The advantages of homogeneous acid catalysts over homogeneous base catalysts are that they are tolerant for water and high FFA in the raw feed[4, 6]. However, the requirement of higher catalytic volume, increased alcohol requirement and high processing temperatures are significant drawbacks. Apart from that, the catalyst is not reusable and by products like glycerol ethers are formed[5, 9].

Enzyme catalysts like lipase are used in an immobilized form to produce biodiesel to prevent contamination of the enzyme. This method is pretty effective but the produced glycerol inhibits the activity of the catalyst thus reducing its reusability [9, 19]. Also the catalyst shows low reactivity in the presence of water and alcohol also affects the immobilization of the enzyme. Another disadvantage is that the storage of the enzyme is difficult as enzymes are sensitive to the environment and involve very high production cost [20, 21].

Heterogeneous acid catalysts and renewable sources have been proven to be effective in converting low grade oils which contains multiple types of impurity and oils with higher acid value to methyl esters [9, 22]. The catalyst has shown good yield up to 96 % and is environment friendly. But the process requires high temperatures and leaching of catalyst into product restricts this process from being a commercial success [2, 5, 9, 20]. Heterogeneous base catalysts like alkali metal oxides and carbonates like MgO and CaO and base zeolites are very sensitive to high FFA. The FFA poisons the active sites and water hydrolyzes the FFA and slows the reaction rate [5, 19]. This makes the production cost higher and the catalysts hard to work with [2].

Heterogeneous solid acid catalysts in particular Heteropolyacids (HPA), are an excellent replacement for liquid acids reducing corrosion problems, highly water tolerant and ecofriendly [23]. Moreover, HPA with keggin structure possesses high thermal stability and high acidity which makes it a super acid and is also reusable [24]. Among the several synthesized HPA's, 12-tungstophosphoric acid (TPA) is found to have very high activity and has been used for catalyzing a number of reactions [24]. TPA has a low catalytic surface area and is highly soluble in water and alcohols, which makes it

IOP Conf. Series: Materials Science and Engineering **574** (2019) 012022 doi:10.1088/1757-899X/574/1/012022

difficult during the recovery of the catalyst [4, 11, 25, 26]. It has been noted that addition of alkali metals such as Potassium (K^+) and Cesium (Cs^+) ions to TPA is carried out in order to form salts with the respective metal ions [27, 28]. Cesium Tungstophosphoric acid (CsTPA) is thus tolerant to polar solvents and has a higher surface area due to its micro porosity and is preferred [29-31]. It has a longer catalytic life than other heterogeneous catalysts. Using Cesium Tungstophosphoric acid in biodiesel production is very effective as direct transesterification occurs and reduces the number of intermediate processes. [4, 24, 25, 32]

The present work focuses on producing biodiesel from WCO with high FFA (around 60 mg KOH/g) using synthesized nano particles of Cesium Tungstophosphoric acid as catalyst and also to study the engine performance and exhaust emission characteristics of the produced biodiesel. The synthesized catalysts were characterized and the process conditions were optimized for maximum yield. Fatty acid composition was determined using GC and the properties were analyzed as per ASTM standards. Reusability studies were performed for analyzing the efficiency of the catalyst for sustained activity.

2. Materials and Methods

2.1 Materials

WCO was procured from street food vendors in Chennai which served fast food. The collected WCO was subjected to pretreatment to remove suspended solid impurities. AV of the WCO sample was tested to be 60.12 mg KOH/g. Tungstophosphoric Acid (99.9 %) and Cesium Carbonate (99%) were purchased from LobaChemie Pvt. Ltd. Methanol (99.8 % purity by weight) was procured from Himedia Chemicals. Potassium Hydroxide (85 %) and Sodium Sulfate (99.9 %) were purchased from Merck Specialties Pvt. Ltd. Standard Supelco 37-component FAME mix was purchased from Sigma-Aldrich.

2.2 Synthesis of Catalyst

The acidic Cesium Tungstophosphoric acid catalyst (CS1TA) was prepared by adding aqueous solution of 1M Cesium Carbonate ($CsCO_3$) to the aqueous solution of 0.8M Tungstophosphoric acid (TPA) by titration method under continuous rapid stirring [28, 33]. The solution formed a milky suspension which was aged for 20 hours at room temperature and evaporated to 100 °C. The white powder was further calcined at a temperature of 300 °C for a time period of 2 hours. Similar procedure was followed to synthesize Cesium Tungstophosphoric acid catalysts (CS2TA) and (CS3TA) by varying the concentrations of Cesium Carbonate (1.5M) and (2.5M) with aqueous solution of Tungstophosphoric acid.

2.3 Characterization Techniques

BRUKER-Alpha FTIR spectrometer with OPUS software using KBr pellet method for transmission measurements was used in the range of 400-4000 cm $^{-1}$ at 24 scan per spectrum at 22 cm $^{-1}$ resolution to record the FT-IR spectra which was translated into a plot of absorbance percentage against wave number.The GE XRD 3003 TT diffractometer system with XRDA31 software was used to analyze X-ray diffraction, peak positions and peak broadening of Cesium Tungstophosphoric acid (CsTPA) catalysts employing Cu Kα radiation of wavelength 0.154178 nm and a scanning angle between 10-70°.

The morphology of samples were obtained from Field Emission Scanning Electron Microscopy (High Resolution) using HITACHI Model S-4800 fitted with an energy dispersive analytical system (EDS, EDAX HORIBA) with a working potential of 15 kV. MalvemZetasizer Nano ZS96 analyzer with a count rate of 291.4 kcps was employed to analyze the particle size distribution in aqueous media by the method of Dynamic Light Scattering. Samples were dispersed using double deionized water at 25°C for 60 seconds in a disposable cuvette.

Micrometrics ASAP 2020 and Porosity analyzer using N_2 adsorption-desorption isotherm method were used to measure the surface area and pore size distribution respectively at a temperature of 77K. The samples were pretreated by degassing under vacuum at 373K for 3 hours. Barrett-Joyner-Halenda (BJT) method was employed for calculating the pore size distribution.

FAME conversion of WCO was analyzed using GC-FID (Young Lin Instruments, 6500GC System) with injector, Flame Ionization Detector (FID) and DB-WAX [polyethylene glycol (PEG)] - Capillary column. The length, diameter and film thickness of the column used were 30 m, 0.25 mm and 0.25 μm, respectively. The injection was executed in split mode with a split ratio of 10:1 at 0.1 minute. The temperature maintained in the injector and detector were 220 °C and 250 °C, respectively while the oven temperature was programmed at 100 °C with 3 minute holding. With a heating rate of 8 °C/min, the temperature was increased to 240 °C for 10 minutes holding time. Flow rate of Nitrogen (carrier gas), Hydrogen and Oxygen were maintained at 10, 30 and 300 mL/min respectively. Supelco 37-component FAME mix in Dichloromethane was the standard used in GC analysis.

Properties of WCO and the esters of WCO were tested using the American Standard Testing Methods (ASTM). An electrical dynamometer coupled, single cylinder, water cooled, four stroke Kirloskar Diesel Engine with an engine power of 5 HP, speed 1500 rpm, bore diameter 0.0875m, stroke length 0.11m and a compression ratio measuring 16:5:1 was used for engine test of biodiesel. The engine was connected with AVS DIGAS 444 gas analyzer for exhaust gas emission studies.

2.4 Transesterification Reaction

Experiments were carried out by adding preheated oil to a mixture of preheated methanol and catalyst at 300rpm. The mixture was stirred for specified time duration and temperature and cooled in a separating funnel until it separated into methyl esters (upper layer) and glycerol (lower phase). After the completion of the reaction, the methyl esters were washed with water to produce clear biodiesel. This procedure was duplicated for optimizing the reaction conditions i.e., Catalyst Weight percentage (1.5-6 %), Molar ratio of WCO to Methanol (1:3-1:21), Reaction Time (3-6 h) and Reaction Temperature (50-75^{0}C).The catalyst was separated and then reused. Methanol was recovered using a rotary evaporator.

3. Results and Discussion

3.1 Characterization Studies of Cesium Tungstophosphoric acid

FT-IR studies of synthesized nanoparticles (CS1TA, CS2TA and CS3TA) were recorded after calcination as shown in Fig.1. Generally, keggin structured heteropoly acids are elucidative in finger print region in FT-IR spectroscopy. By analysis of individual bands in the finger print region, it is possible to realize that four oxygen atoms make four characteristic bands in the 1200-700 cm^{-1} range. Appearance of the bands is based on the degree of hydration as well as the nature of counter cations present [34]. In keggin structured heteropolyacids, PO_4 tetrahedron is

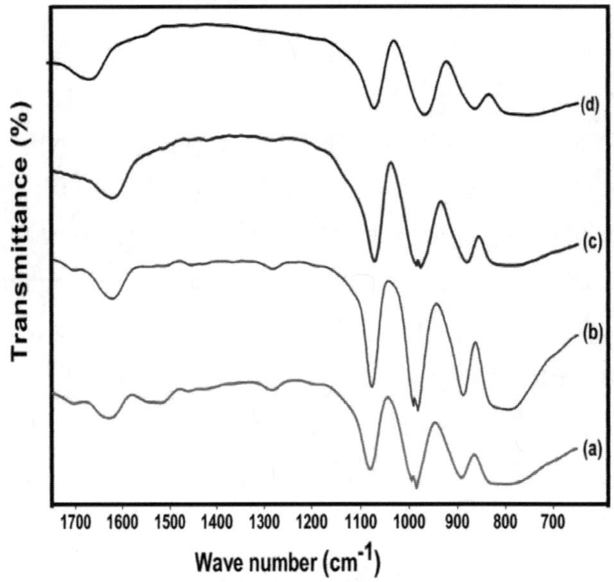

Fig.1: FT-IR Spectra of (a) CS1TA (b) CS2TA (c) CS3TA and (d) TPA

surrounded with 12 WO_6 octahedral with shared edges in W_3O_{13} triad group and corner between each triad via oxygen atom. Typical vibration peaks of keggin structured tungstophosphoric acid showed strong absorbance bands as: 1081 cm^{-1} (Stretching band of P-O in the central tetrahedron of PO_4), 983 cm^{-1} (W=O_f terminal oxygen in the exterior WO_6 octahedron), 796 (bands for the W-O_c-W corner sharing) and 893 cm^{-1} (bands for the W-O_b-W bridge sharing) which coincide with literature [28]. The characteristic vibration of W=O at 983 cm^{-1} split into 2 peaks at 985 and 993 cm^{-1} respectively which can be accounted for W=O with H^+ (presence of H_2O) and interaction with Cs^+ ion. These results confirm the formation of CsTPA from Tungstophosphoric acid and Cesium carbonate.

XRD patterns of synthesized CS1TA, CS2TA, CS3TA and bulk TPA in the 2θ angle range between 10-70° is shown in Fig.2. The five reflection peaks between 25° to 40° can be accounted for the keggin structure of TPA. It can be seen that the location and crystalline structure of all the synthesized samples are very similar to bulk TPA. But the peaks are slightly inclined towards the higher angle in CS1TA, CS2TA and CS3TA which can be attributed to the presence of BCC structure of Cesium tungstophosphoric acid [32]. The XRD patterns of the crystalline phases in the synthesized samples were analysed and compared with JCPDS (Joint Committee of the Powder Diffraction

Standards) and the peaks confirmed the presence of keggin structured Tungstophosphoric acid in the synthesized samples (JCPDS No. 50-0305).

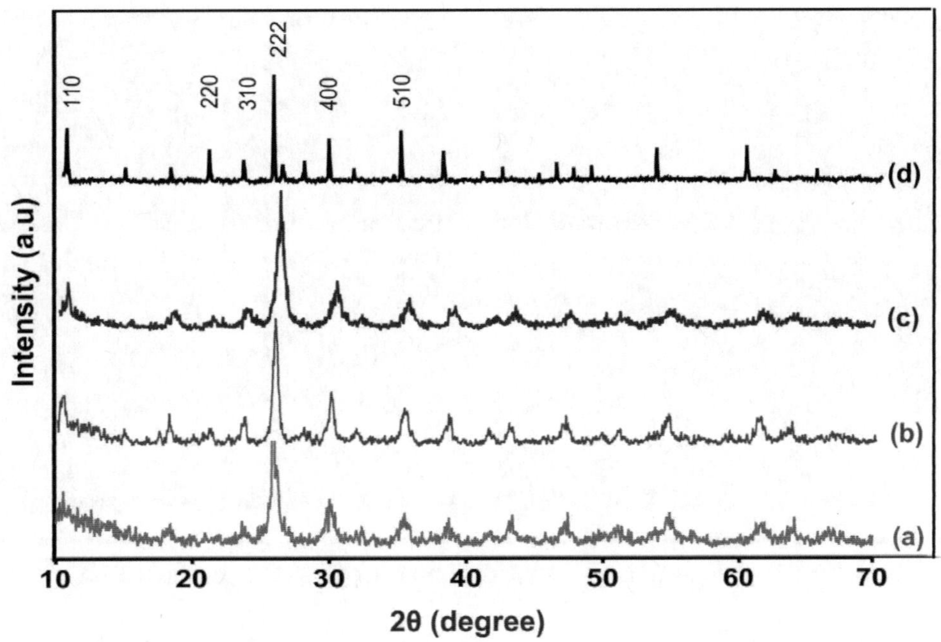

Fig.2: Powder XRD patterns of (a) CS1TA (b) CS2TA (c) CS3TA and (d) TPA

The average crystalline size of synthesized cesium tungstophosphoric acid and bulk TPA were also determined with the help of most intense peak using Debye-Scherrer's formula. Average crystalline size of CS1TA, CS2TA and CS3TA and bulk TPA were found to be 18.45, 13.28, 9.76 and 56.59 nm respectively. The X-ray diffraction studies show that CS3TA was found to have the lowest crystalline size when compared to CS1TA, CS2TA and bulk TPA.

Surface morphology studies of the synthesized catalysts CS1TA, CS2TA, CS3TA and bulk TPAusing high resolution SEM, calcined at N_2 atmosphere is given in Fig.3(a-d). Fig.3 a&b depicts the SEM images of synthesized samples CS1TA and CS2TA which are spherically aggregate and have size below 100 nm.CS3TA was found to be spherical in nature with well-defined crystalline particle size less than 50 nm as shown in Fig.3c. SEM images of TPA as shown in Fig.3d shows particles with irregular shape. EDX spectra also recorded at various positions of samples indicate the presence of elements Cesium, Oxygen, Phosphorus and Tungsten in all the synthesized samples as shown in Table 1.

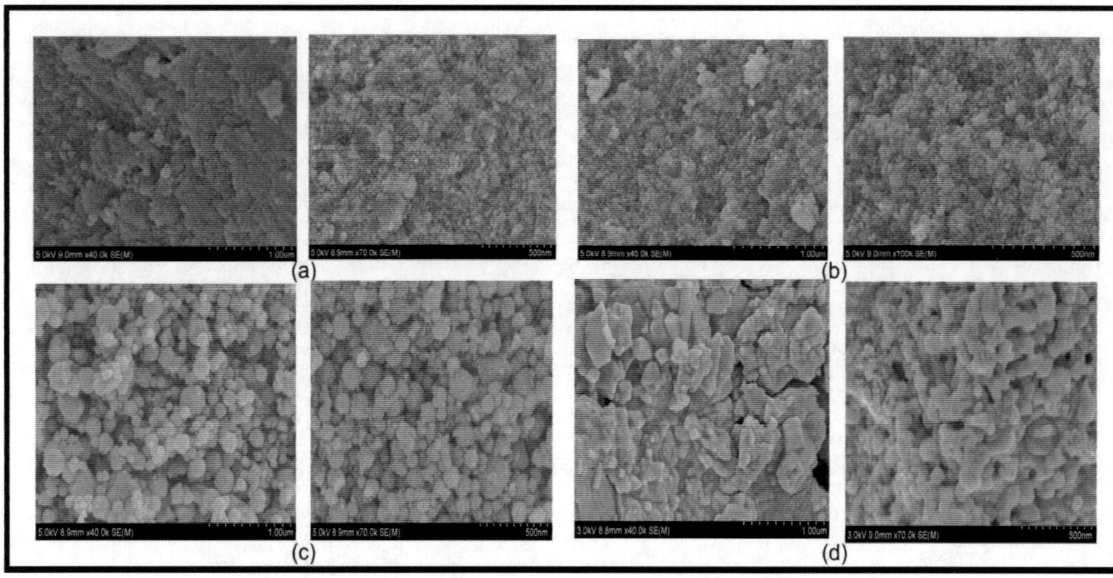

Fig.3: SEM images of (a) CS1TA (b) CS2TA (c) CS3TA and (d) TPA

Table.1: EDX quantification elements (a) CS1TA (b) CS2TA (c) CS3TA and (d) TPA

Catalysts	Elements							
	% of Cesium		% of Phosphorus		% of Oxygen		% of Tungsten	
	Atomic	Weight	Atomic	Weight	Atomic	Weight	Atomic	Weight
CS1TA	5.24	12.07	2.02	1.21	73.53	20.37	21.22	67.56
CS2TA	4.42	10.03	2.19	1.19	73.31	20.03	22.27	69.94
CS3TA	4.05	9.90	2.33	1.33	73.79	21.71	19.83	67.06
TPA	-	-	2.01	1.37	79.29	26.18	18.71	73.82

Dynamic Light Scattering (DLS) method measures the size as well as the Z-average of the nano particles using light scattered from a laser beam which passes through a colloidal suspension. The samples CS1TA, CS2TA, CS3TA and bulk TPA were dispersed in double deionized water to make a colloidal solution. The particle size and Z-average of bulk TPA was found to be 194 and 438 d.nm, respectively. Particle size of the catalysts increased from 194 to 506 d.nm when substituting TPA with Cesium ions. Particle size of catalysts CS1TA, CS2TA and CS3TA was found to be 216, 239 and 506 d.nm respectively as shown in Fig.4. On the other hand, when decreasing the percentage of TPA, the Z-average of catalysts decreased from 438 to 154 d.nm. The Z-average of catalyst CS1TA, CS2TA and CS3TA was found to be 368, 160 and 154 d.nm respectively. Among the synthesized catalysts, CS3TA was found to have the lowest Z-average and highest particle size.

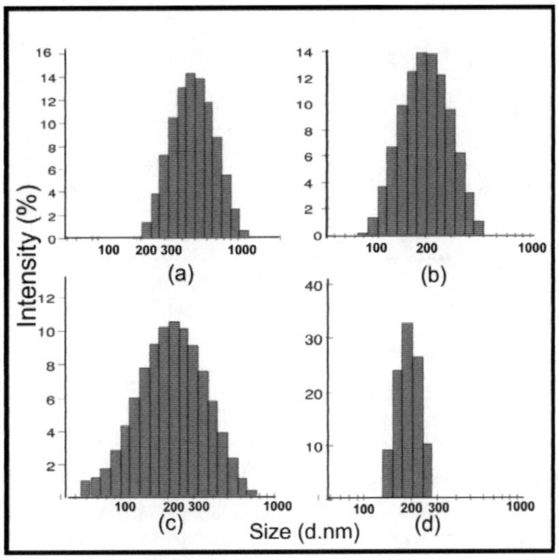

Fig.4: Average Particle Size Distribution of (a) CS1TA (b) CS2TA (c) CS3TA and (d) TPA

Surface area and pore size distribution studies using N_2 adsorption-desorption isotherms for all the synthesized catalysts are shown in Fig.5. Single and multi-point BET surface area, pore size and pore volumes of synthesised catalysts are given in Table 2. The single point surface area of catalysts CS1TA, CS2TA and CS3TA were found to be 3.75, 23.39 and 94.86 m^2/g, respectively. The multi-point BET surface area of catalysts CS1TA, CS2TA and CS3TA were found to be 5.81, 23.39 and 96.50 m^2/g, respectively.Since surface area is inversely proportional to pore size, we see from Table 2, that the pore size of the catalysts decreased with increase in surface area. On the other hand, the pore volume of the catalyst increased with increase in surface area.

Table-2:Single and multipoint BET, average pore size and pore volumes of synthesized catalysts (a) CS1TA (b) CS2TA (c) CS3TA

Samples	Surface area (m^2/g)		Pore size $d_{(BET)}$ (nm)	Pore volume V_{BJH} (cm^3/g)
	Single point (BET)	Multi point (BET)		
CS1TA	3.75	5.81	12.41	0.0118
CS2TA	23.39	23.26	3.23	0.0187
CS3TA	94.86	96.50	4.82	0.1164

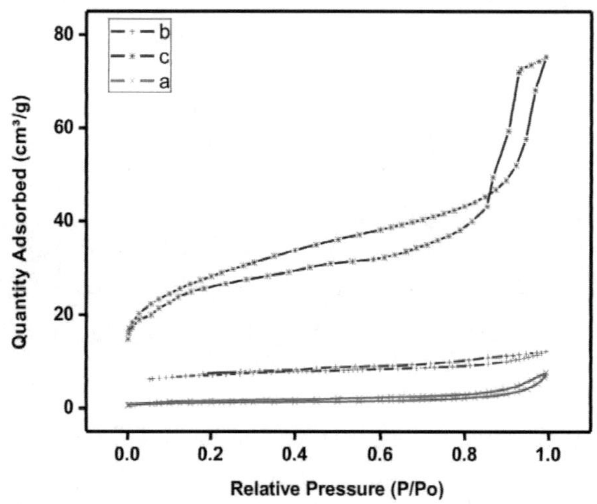

Fig.5: N_2 Adsorption-Desorption of (a) CS1TA (b) CS2TA and (c) CS3TA

Based on the characterization studies, catalyst CS3TA has high surface area, lower crystalline size, increased particle size and lower Z-average when compared to other catalysts and hence used for synthesis of biodiesel from WCO with high AV.

3.2 Optimization Studies of Process Variables

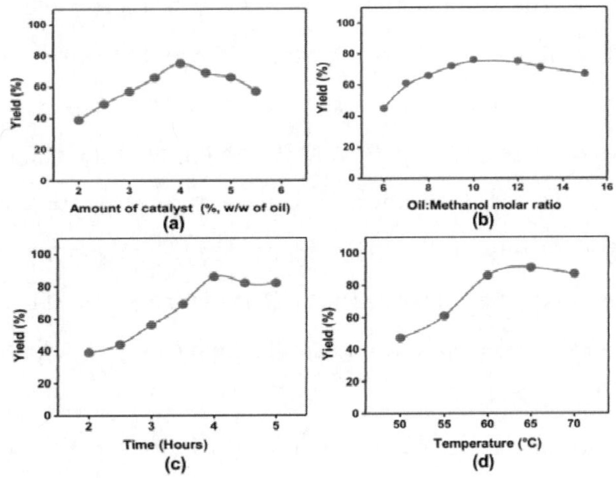

Fig.6: (a) Effect of catalyst Amount (b) Effect of methanol to oil molar ratio (c) Effect of duration of the reaction and (d) Effect of temperature of the reaction

Experiments were carried out by varying the amount of the catalyst in the range 2% to 5.5% (w/w of oil) while keeping the other reaction parameters constant at 9:1 alcohol to oil molar ratio for 5 hours at 60°C (Fig.6a). The yield increased from 39 to 75 %, with the increase in catalyst amount from 2 to 4% (w/w of oil). Increase in the yield could be attributed to the presence of more active sites for the transesterification reaction to take place (34). However, when the amount of catalyst used for the conversion was increased to quantity higher than 4 % (w/w of oil), decrease in yield of FAME was noted. The decrease in yield may be due to the excess slurry formation making the mixture too viscous [35].

Due to the reversible and exothermic nature of the process, the molar ratio of alcohol was varied from 6:1 to 15:1 at constant catalyst concentration 4% (w/w of oil), 5 hours at 60 °C to get maximum yield. Yield percentage of biodiesel from WCO increased from 45 % to 75 % with increase in alcohol ratio from 6:1 to 10:1 as shown in Fig.6b. The yield % decreased beyond 10:1 which could be due to the increase in viscosity due to the accumulation of excess methanol in the mixture [36].

Effect of reaction time on FAME yield from WCO was studied by varying the time from 2-5 hours at a constant molar ratio of methanol to oil (10:1) and catalyst amount 4% (w/w of oil) at 60°C. It can be seen that the biodiesel yield increased to 86% with increase in reaction time as shown in

IOP Conf. Series: Materials Science and Engineering **574** (2019) 012022 doi:10.1088/1757-899X/574/1/012022

Fig.6c. Beyond 4 hours, the yield decreased slightly due to the blockage of pore openings of the catalyst with increase in time. Therefore, 4 hours was selected as appropriate transesterification time.

In order to study the influence of temperature on the yield, it was varied from 50 to 70°C and the effect on the transesterification process was studied as shown in Fig. 6d. With increase in temperature, the FAME yield increased to 91% at 65°C which might be due to the greater miscibility of methanol with oil and enhanced diffusion rate. Further increase in temperature to 70°C slightly decreased the yield from 91% to 84.9 %. This may be due to the vaporization of methanol which resists the reaction on the three phase interface [36]. The synthesized catalyst exhibited appreciable conversion of FAME even at low temperatures due to the high catalytic activity.

3.3 Reusable Studies of the catalyst

A remarkable feature of the heteropolyacid catalyst is its stability and sustained activity on repeated use. The catalyst was separated at the end of the reaction, stirred with methanol and hexane for 2 hours. It was dried at 120°C for 2 h to remove traces of moisture and reused. The results show a decrease in biodiesel yield by 4 to 16 % on recycling up to five cycles (Fig.7). The catalyst showed stable activity even after recycling.

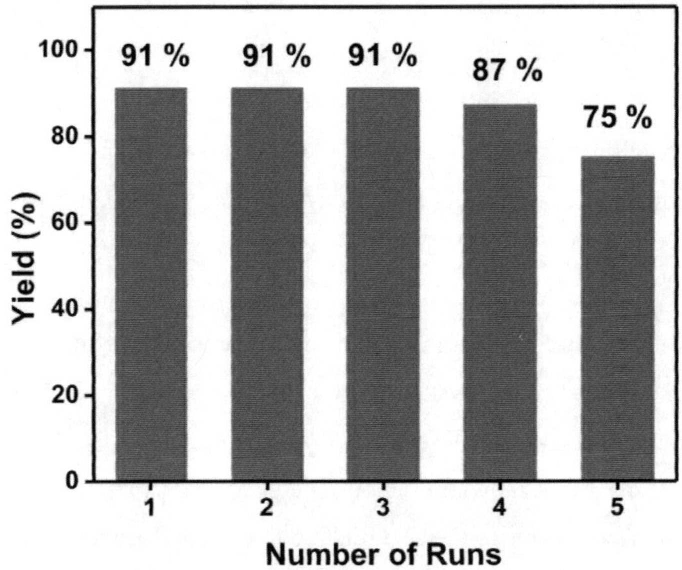

Fig.7: Reusability of catalyst CS3TA

3.4 Biodiesel properties

The important quality parameters of FAME obtained from WCO and raw WCO are shown in Table 3. All the measured values were in accordance with ASTM standards. The Cetane Number(CN) of FAME is found to be more than diesel. The cooking process and the molecular structure of WCO have an impact on the cetane number and other properties. Low-sulfur fuels play a vital role in reducing exhaust emissions. It can be seen that the sulfur content in the biodiesel is very low when compared with diesel, thus contributing to reduced emissions.

Table 3:Properties of Waste cooking oil, Biodiesel and Diesel

Properties	Unit	Method	WCO	WCOME	DIESEL
Acid Value	mg of KOH/g	ASTMD974-14e2	60.32	0.81	NA
Density	kg/m^3 (30 °C)	ASTM D1298	913	818	793
Kinematic Viscosity	mm^2/S, 40 °C	ASTM D4016-14	5.4	4.8	3.5
Calorific value	MJ/kg	ASTM D240	35.56	43.91	41.95
Water content	% w/w	ASTM D2709	0.05	0.01	0.01
Cloud point	° C	ASTM D5772	+8	+6	-14
Pour point	° C	ASTM D97	+5	+3	-18
Flash point	° C	ASTM D93	208	124	75
Fire point	° C	ASTM D9390	220	133	82
Sulfur content	Mass %	ASTM D5453	0.01	0.01	0.95
Copper strip corrosion	50 °C, 3 hours	ASTM D130-12	1	1	1
Iodine value	$g\ I_2\ 100\ g^{-1}$	ASTM D5768	72.5	96.2	-
Cetane index	-	ASTM D976-06	42	53	49

3.5 Gas Chromatography analysis

The peaks in the WCOME chromatogram correspond to the presence of various esters in the sample and were identified by comparing with the standard FAME mixture. The compositions of FAME present in WCOME are shown in table-4 and Fig.8.

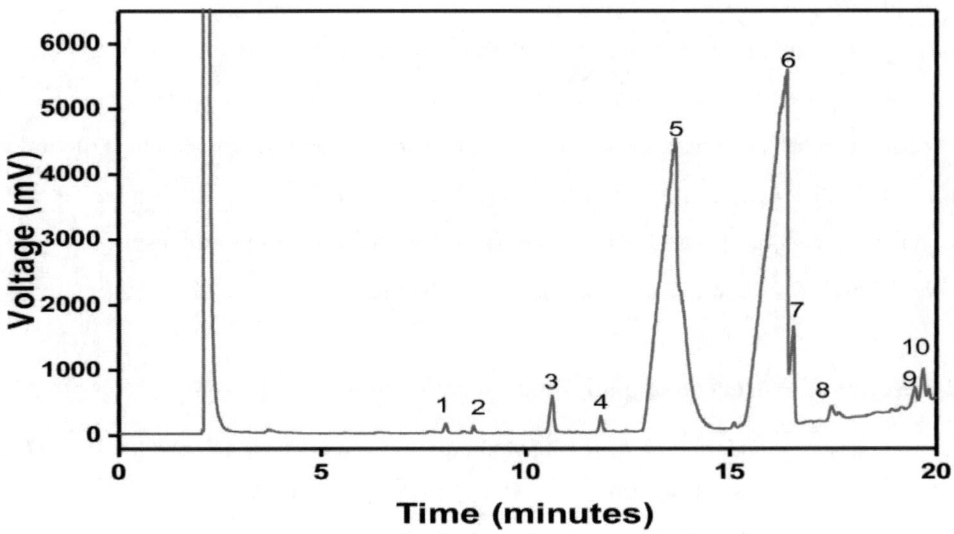

Fig.8: GC chromatogram of WCOME

Table-4: Composition of FAME obtained from WCO

Peaks	Area (%)	Identified esters
1	0.13	Methyl dodecanate
2	0.05	Methyl tetradecanate
3	1.05	Methyl pentadecanate
4	0.25	Methyl octadecanate
5	42.0	Methyl 9-octadecenate
6	53.9	Methyl hexadecanate
7	0.49	Methyl 9-hexadecenate
8	0.25	Methyl 9,12-octadecadienate
9	0.41	Methyl eicosanate
10	0.86	Methyl 11-eicosenate

3.6 Performance Studies

Brake Thermal Efficiency (BTE), Mechanical Efficiency (ME) and Brake Specific Fuel Consumption (BSFC) of biodiesel obtained from WCO and diesel were investigated.

Variation of BTE with respect to load percentages for diesel and WCOME is presented in Fig.9a. The BTE of an engine is inversely related to the fuel consumption. Results show an increase of BTE for WCOME and diesel with increase in load percentages. Better fuel consumption of both WCOME and diesel, leads to proper atomization, resulting in increase in power and reduction in heat loss. Lower BTE of WCOME is attributed to higher fuel consumption and low calorific value when compared to diesel [37]. The lesser heat content and the unsaturated condition of WCOME is another reason for reduction in BTE.

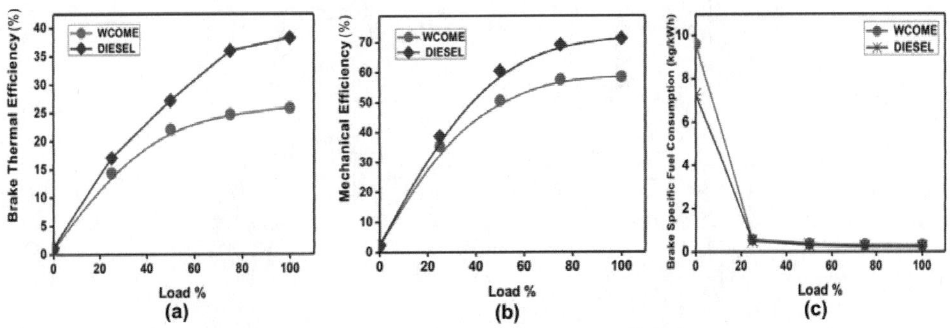

Fig.9: (a) BTE vs Load % (b) ME vs Load % and (c) BSFC vs Load %

The Mechanical Efficiency (ME) of WCOME and diesel were investigated with respect to load percentages as shown in Fig.9b. ME increased with increase in load percentages for both WCOME and diesel. But the mechanical efficiency of WCOME was slightly lower due to higher density, kinematic viscosity and low volatility which could have affected the atomization of the fuel. At 25 % load, the ME of WCOME was close to diesel but further increase in load % increased the ME for both WCOME and diesel. The average decrease in ME of WCOME was found BSFC of engine operated with both WCOME and diesel at various engine load percentages are given in Fig.9c. BSFC is inversely related to the BTE and has a tendency to decrease with increasing load percentages. Properties like kinematic viscosity, density, heating value and fuel injection influence the BSFC. The higher density of biodiesel leads to the plunger discharging more fuel in the fuel injection pump which eventually leads to increase in difference of BSFC between WCOME and diesel. BSFC of both WCOME and diesel decreased by increasing the load percentages [38]. At 25 % of load, the variation between WCOME and diesel was found to be 13.8 %. Further increase in load conditions *viz,* 50 %, 75 % and 100 %, increased the variation of BSFC between WCOME and diesel by 17.8 %, 27.9 % and 29.3 %, respectively.

3.7 Emission Studies

Emission of Carbon monoxide (CO), Carbon dioxide (CO_2), unburned Hydrocarbon (HC) and Oxides of Nitrogen were investigated for both diesel and biodiesel obtained from WCO.

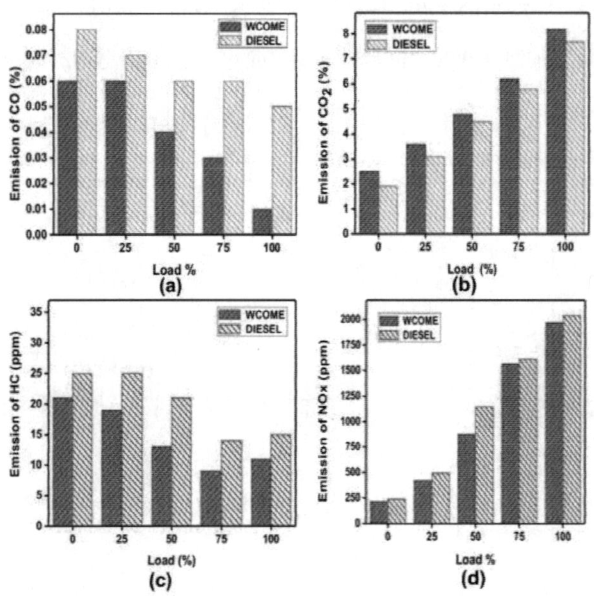

Fig.10: (a) CO emission vs Load % (b) CO_2 emission vs Load % (c) HC emission vs Load % and (d) NOx emission vs Load %

Formation of CO is due to the partial oxidation from CO to CO_2 and insufficient burning of carbon atoms in the fuel. Emission of WCOME decreased for all load conditions with respect to diesel as given in Fig.10a. Concentration of oxygen in the fuel, engine speed and mixing of air and fuel are the influencing factors affecting the emission of CO [39]. The possibility of forming a fuel-rich zone is reduced due to the higher CN of biodiesel thus resulting in reduced CO emissions [40].

Variation of CO_2 emission of WCOME and diesel is shown in Fig. 10b. It is found that the emission of both WCOME and diesel increased with increasing load percentages. The high

IOP Conf. Series: Materials Science and Engineering **574** (2019) 012022 doi:10.1088/1757-899X/574/1/012022

concentration of CO_2 emission of FAME when compared with diesel is mainly attributed to the presence of high carbon and oxygen leading to complete combustion of biodiesel [41].

Fig. 10c depicts the variation of HC emissions for FAME and diesel from WCO. It is clear from the figure that, when the diesel engine is run on pure biodiesel the HC emissions are reduced [42]. The reason behind this reduction may be because of the increase in chain length of biodiesel (28 %) when compared to diesel [43]. CN also plays an important role in reducing the HC emissions. As the load percentage is increased, there is a considerable reduction in HC emission for WCOME due to higher CN leading to burning delay of fuel [44].

Emission of NO_x is a major concern in biodiesel synthesis. This is mainly due to combustion temperature, time and oxygen present in the fuel. Fig.10d shows the variation of NO_x emission levels with varying load percentages. It is found that the NOx for both WCOME and diesel increased with increasing load percentages which might be due to increase in combustion temperatures and pressure inside the cylinder [45]. However, the concentration of NOx was lower at all load percentages for WCOME than diesel with 23.5 % reduction at 50 % load [46]. This reduction in NOx could be due to the catalytic activity of the synthesized heteropolyacid. It was encouraging to find an average reduction of 10.3 % when compared with diesel.

Conclusion

Catalytic activity of synthesized CsTPA nanoparticles for transesterifying high fatty acid WCO into biodiesel with lower NOx and HC emissions was studied. The synthesized catalysts were characterized using FTIR, XRD, DLS, BET and SEM. CS3TA with high surface area (96.50 m^2/g), lower crystalline size (9.76 nm) and lower Z-average (154 d.nm) was used for producing FAME from WCO. Reaction Parameters (temperature, alcohol/oil molar ratio, catalyst quantity and time) for maximizing the yield of WCOME were investigated. Highest yield of biodiesel was obtained at 65 °C, 4 hours at catalyst amount 4 % (w/w of oil) and 10:1 alcohol/oil molar ratio. FAME produced was in compliance with the ASTM standards. It was found that BTE and ME values were low for biodiesel than diesel. The average decrease in ME of WCOME was found to be 13.7 % when compared to diesel. HC emissions decreased by 28% in comparison with the conventional fuel. An average reduction in NOx emission by 10.3 % was due to the catalytic activity of the synthesized heteropolyacid. Thus Cesium Tungstophosphoric acid, proved to be a potential reusable catalyst for producing eco-friendly biodiesel with improved performance and lower emissions.

Acknowledgement

The authors thank, Department of Science and Technology, Government of India for funding this research work.

References

[1] A.L. Ahmad, N.H.M. Yasin, C.J.C. Derek, J.K. Lim, Microalgae as a sustainable energy source for biodiesel production: A review, Renewable and Sustainable Energy Reviews 15(1) (2011) 584-593.

[2] D.Y.C. Leung, X. Wu, M.K.H. Leung, A review on biodiesel production using catalyzed transesterification, Applied Energy 87(4) (2010) 1083-1095.

[3] A.E. Atabani, Silitonga, A. S., Badruddin, Irfan Anjum, Mahlia, T. M. I., Masjuki, H. H., Mekhilef, S., A comprehensive review on biodiesel as an alternative energy resource and its characteristics, Renewable and Sustainable Energy Reviews 16(4) (2012) 2070-2093.

[4] F. Su, Y. Guo, Advancements in solid acid catalysts for biodiesel production, Green Chemistry 16(6) (2014) 2934-2957.

[5] M.K. Lam, K.T. Lee, A.R. Mohamed, Homogeneous, heterogeneous and enzymatic catalysis for transesterification of high free fatty acid oil (waste cooking oil) to biodiesel: A review, Biotechnology Advances 28(4) (2010) 500-518.

[6] L.L.N.G. Angelo C. Pinto, Michelle J. C. Rezende, Núbia M. Ribeiro, Ednildo A.Torres, Wilson A. Lopes, Pedro A. de P. Pereira and Jailson B. de Andrade, Biodiesel: An Overview, J. Braz. Chem. Soc. 16 (2005).

[7] Sahar, S. Sadaf, J. Iqbal, I. Ullah, H.N. Bhatti, S. Nouren, R. Habib ur, J. Nisar, M. Iqbal, Biodiesel production from waste cooking oil: An efficient technique to convert waste into biodiesel, Sustainable Cities and Society 41 (2018) 220-226.

[8] M.G. Kulkarni, A.K. Dalai, Waste Cooking Oil-An Economical Source for Biodiesel: A Review, Industrial & Engineering Chemistry Research 45(9) (2006) 2901-2913.

[9] A. Talebian-Kiakalaieh, N.A.S. Amin, H. Mazaheri, A review on novel processes of biodiesel production from waste cooking oil, Applied Energy 104(Supplement C) (2013) 683-710.

[10] L.P. Wang Yong, Ou Shiyi, Zhang Zhisen, Preparation of biodiesel from waste cooking oil via two-step catalyzed process, Energy Conversion and Management 48(1) (2007) 184-188.

[11] A. Anitha, Biodiesel: An Alternate Fuel from Waste Cooking Groundnut Oil Using Supported Heteropolyacids for Green Earth, in: M. Ramkumar (Ed.), On a Sustainable Future of the Earth's Natural Resources, Springer Berlin Heidelberg, Berlin, Heidelberg, 2013, pp. 467-475.

[12] M.C. Math, S.P. Kumar, S.V. Chetty, Technologies for biodiesel production from used cooking oil — A review, Energy for Sustainable Development 14(4) (2010) 339-345.

[13] R. Luque, J.C. Lovett, B. Datta, J. Clancy, J.M. Campelo, A.A. Romero, Biodiesel as feasible petrol fuel replacement: a multidisciplinary overview, Energy & Environmental Science 3(11) (2010) 1706-1721.

[14] Kulkarni, M. G., Dalai, A. K., Waste Cooking OilAn Economical Source for Biodiesel: A Review, Industrial & Engineering Chemistry Research 45(9) (2006) 2901-2913.

[15] I. Nurfitri, G.P. Maniam, N. Hindryawati, M.M. Yusoff, S. Ganesan, Potential of feedstock and catalysts from waste in biodiesel preparation: A review, Energy Conversion and Management 74(Supplement C) (2013) 395-402.

[16] Z. Al-Hamamre, J. Yamin, Parametric study of the alkali catalyzed transesterification of waste frying oil for Biodiesel production, Energy Conversion and Management 79 (2014) 246-254.

[17] J.V. Gerpen, Biodiesel processing and production, Fuel Processing Technology 86(10) (2005) 1097-1107.

[18] J.F.G. Oliveira, I.L. Lucena, R.M.A. Saboya, M.L. Rodrigues, A.E.B. Torres, F.A.N. Fernandes, C.L. Cavalcante, E.J.S. Parente, Biodiesel production from waste coconut oil by esterification with ethanol: The effect of water removal by adsorption, Renewable Energy 35(11) (2010) 2581-2584.

[19] I.M. Atadashi, M.K. Aroua, A.R. Abdul Aziz, N.M.N. Sulaiman, Production of biodiesel using high free fatty acid feedstocks, Renewable and Sustainable Energy Reviews 16(5) (2012) 3275-3285.

[20] A.P. Vyas, J.L. Verma, N. Subrahmanyam, A review on FAME production processes, Fuel 89(1) (2010) 1-9.

[21] K.F. Haigh, G.T. Vladisavljević, J.C. Reynolds, Z. Nagy, B. Saha, Kinetics of the pre-treatment of used cooking oil using Novozyme 435 for biodiesel production, Chemical Engineering Research and Design 92(4) (2014) 713-719.

[22] R. Shan, L. Lu, Y. Shi, H. Yuan, J. Shi, Catalysts from renewable resources for biodiesel production, Energy Conversion and Management 178 (2018) 277-289.

[23] S.A.Basheer. A. Anitha, Comparative Performance and Emission Characteristics of Methyl Esters from Waste Cooking Oil and Used Palm Oil using a Solid Acid Catalyst, Pollution Research 37(4) (2018) 1084-1092.

[24] A. Anitha, Heteropolyacids as heterogenous and reusable catalysts for trans-esterification reactions, Discovery Sci. 1(1) (2012) 14-18.

[25] P. Morin, B. Hamad, G. Sapaly, M.G. Carneiro Rocha, P.G. Pries de Oliveira, W.A. Gonzalez, E. Andrade Sales, N. Essayem, Transesterification of rapeseed oil with ethanolI. Catalysis with homogeneous Keggin heteropolyacids, Applied Catalysis A: General 330(Supplement C) (2007) 69-76.

[26] Z. Yuan, S. Xia, P. Chen, Z. Hou, X. Zheng, Etherification of Biodiesel-Based Glycerol with Bioethanol over Tungstophosphoric Acid To Synthesize Glyceryl Ethers, Energy & Fuels 25(7) (2011) 3186-3191.

[27] I. Holclajtner-Antunović, U.B. Mioč, M. Todorović, Z. Jovanović, M. Davidović, D. Bajuk-Bogdanović, Z. Laušević, Characterization of potassium salts of 12-tungstophosphoric acid, Materials Research Bulletin 45(11) (2010) 1679-1684.

[28] J.A. Dias, E. Caliman, S.C. Loureiro Dias, Effects of cesium ion exchange on acidity of 12-tungstophosphoric acid, Microporous and Mesoporous Materials 76(1) (2004) 221-232.

[29] M.K. Toshio Okuhara, Teruyuki Nakato, A water-tolerant solid acid, CS2.sHo.sPW12040, for hydrolysis of esters in water, Applied Catalysis A: General 155 (1997) L9-L13.

[30] H. Doğan, T.Y. Inan, E. Unveren, M. Kaya, Effect of cesium salt of tungstophosphoric acid (Cs-TPA) on the properties of sulfonated polyether ether ketone (SPEEK) composite membranes for fuel cell applications, International Journal of Hydrogen Energy 35(15) (2010) 7784-7795.

[31] K. Narasimharao, D.R. Brown, A.F. Lee, A.D. Newman, P.F. Siril, S.J. Tavener, K. Wilson, Structure–activity relations in Cs-doped heteropolyacid catalysts for biodiesel production, Journal of Catalysis 248(2) (2007) 226-234.

[32] K. Srilatha, R. Sree, B.L.A. Prabhavathi Devi, P.S. Sai Prasad, R.B.N. Prasad, N. Lingaiah, Preparation of biodiesel from rice bran fatty acids catalyzed by heterogeneous cesium-exchanged 12-tungstophosphoric acids, Bioresource Technology 116(Supplement C) (2012) 53-57.

[33] M. Kimura, T. Nakato, T. Okuhara, Water-tolerant solid acid catalysis of Cs2.5H0.5PW12O40 for hydrolysis of esters in the presence of excess water, Applied Catalysis A: General 165(1) (1997) 227-240.

[34] N. Essayem, A. Holmqvist, P.Y. Gayraud, J.C. Vedrine, Y. Ben Taarit, In Situ FTIR Studies of the Protonic Sites of H3PW12O40 and Its Acidic Cesium Salts MxH3−xPW12O40, Journal of Catalysis 197(2) (2001) 273-280.

[35] Y. Tang, J. Xu, J. Zhang, Y. Lu, Biodiesel production from vegetable oil by using modified CaO as solid basic catalysts, Journal of Cleaner Production 42 (2013) 198-203.

[36] L.C. Meher, D. Vidya Sagar, S.N. Naik, Technical aspects of biodiesel production by transesterification—a review, Renewable and Sustainable Energy Reviews 10(3) (2006) 248-268.

[37] N.R. Banapurmath, P.G. Tewari, R.S. Hosmath, Performance and emission characteristics of a DI compression ignition engine operated on Honge, Jatropha and sesame oil methyl esters, Renewable Energy 33(9) (2008) 1982-1988.

[38] S. Godiganur, C. Suryanarayana Murthy, R.P. Reddy, Performance and emission characteristics of a Kirloskar HA394 diesel engine operated on fish oil methyl esters, Renewable Energy 35(2) (2010) 355-359.

[39] C.-Y. Lin, H.-A. Lin, Diesel engine performance and emission characteristics of biodiesel produced by the peroxidation process, Fuel 85(3) (2006) 298-305.

[40] F. Wu, J. Wang, W. Chen, S. Shuai, A study on emission performance of a diesel engine fueled with five typical methyl ester biodiesels, Atmospheric Environment 43(7) (2009) 1481-1485.

[41] P.K. Sahoo, L.M. Das, M.K.G. Babu, S.N. Naik, Biodiesel development from high acid value polanga seed oil and performance evaluation in a CI engine, Fuel 86(3) (2007) 448-454.

[42] B.-F. Lin, J.-H. Huang, D.-Y. Huang, Experimental study of the effects of vegetable oil methyl ester on DI diesel engine performance characteristics and pollutant emissions, Fuel 88(9) (2009) 1779-1785.

[43] G. Knothe, C.A. Sharp, T.W. Ryan, Exhaust Emissions of Biodiesel, Petrodiesel, Neat Methyl Esters, and Alkanes in a New Technology Engine, Energy & Fuels 20(1) (2006) 403-408.

[44] G.H. Abd-Alla, H.A. Soliman, O.A. Badr, M.F. Abd-Rabbo, Effects of diluent admissions and intake air temperature in exhaust gas recirculation on the emissions of an indirect injection dual fuel engine, Energy Conversion and Management 42(8) (2001) 1033-1045.

[45] S. Puhan, N. Vedaraman, G. Sankaranarayanan, B.V.B. Ram, Performance and emission study of Mahua oil (madhuca indica oil) ethyl ester in a 4-stroke natural aspirated direct injection diesel engine, Renewable Energy 30(8) (2005) 1269-1278.
[46] H. Aydin, H. Bayindir, Performance and emission analysis of cottonseed oil methyl ester in a diesel engine, Renewable Energy 35(3) (2010) 588-592.

Investigation on Mode I Interlaminar Fracture Toughness of Chemically Treated / Untreated Saw Dust Powder Based Jute Fabric Reinforced Epoxy Composite Structure

R. Suthan[a*], V. Jayakumar[b], S. Madhu[c]
[a]Department of Mechanical Engineering, Saveetha School of Engineering, Saveetha Institute of Medical and Technical Sciences, Chennai – 602 105, Tamil Nadu, India.
[b]Department of Mechanical Engineering, Amrita School of Engineering, Amrita Vishwa Vidyapeetham, Chennai Campus, Chennai – 601 103, Tamil Nadu, India.
[c]Department of Automobile Engineering, Saveetha School of Engineering, Saveetha Institute of Medical and Technical Sciences, Chennai – 602 105, Tamil Nadu, India.

Corresponding Author: itsmesuthan@gmail.com

Abstract. Nowadayspolymer matrix composites are widely used in the areas of automotive, medical, aircraft and aerospace industries. Since, chemically surface modified saw dust with jute fiber reinforced composites play vital role in developing lightweight structural materials in many engineering industries. This study focuses on utilizingchemically surface modified saw dust with jute fiberas reinforcement filler loading malic acid treated and un treated (10 %,20% and 30%) and fiber loading (3-layer jute fiber) in epoxy resin. This composite was fabricated using hand lay-up process. The influence of filler volume fraction and fiber weight fraction on the fracture toughness properties of treated / untreated modified saw dust with jute fiber/filler was studied initially. Test results revealed that the composite 20-volume fraction malic acid treated saw dust with 3-layer jute fiber composite has better fracture properties as compared with other specimens.
Key words: Epoxy, chemically surface modified saw dust with jute fiber, fracture properties.

1. Introduction

Koktaet. al (1989) studied the mechanical behaviour of wood flour - added polypropylene composite and found the addition of filler material in the polypropylene matrix increases the mechanical properties of the composite material. Sarki et al(2011) have used coconut shell powder as filler material in epoxy composite and it was found that incorporation of coconutshell powder increased the tensile strength and modulus with a slight decrease in impact strength. Mohamad Hana

IOP Conf. Series: Materials Science and Engineering **574** (2019) 012023 doi:10.1088/1757-899X/574/1/012023

et al (2017) investigated the coconutshell powder with PVC composite for automobile application. This research revealed that tensile, flexural and impact strength as well as it modulus of elasticity is improved by 42%, 25% and 23% respectively when compared to the pure resin. Chun et al (2013) used coconut shell powder as filler in recycled polypropylene, and sodium dedecyl sulfate (SDS) as couplingagent. It revealed that the addition of filler provide danin crease intensile properties, thermal stability, and crystalinity and lower the water absorption compared to unmodified composites. Kuburi et al (2017) investigated the coconut shell powder filled composite prepared from recycled low density polyurethane polymer matrix contain upto 15wt% of coconut shell powder as filler the effect of coir fiber loading and the process of the composite was investigated. The result reveals that both mechanical properties and waster absorption properties are improved. Siro(2010) prepared a completely bio composite based on Poly Lactic Acid(PLA) and olivepit powder and it was reported that addition of filler resulted in increase tensile modulus and decrease inflexural strength. Muthukumar et al(2006) investigate about developing the polymer matrix composite using coconut shell powder and groundnut shell powder in different volume fraction the evaluation showed that the tensile strength, impact strength and flexural property and hydrophilic behaviour has increased. The material properties are greatly influenced by the coconutshell powder and groundnut shell powder volume fraction. Udhaya Sankar et al. (2015) researched the coconut shell powder reinforced polymer composite. Results revealed that the filler addition improves the mechanical properties like tensile strength and impact strength. Chandramohan et al(2017) studied the bio particle composite from coconutshell powder,walnutshell and risehuskin epoxy matrix. The result revealed that 20vol% of filler provides better mechanical property and modulus due to uniform dispersion of filler and matrix. Onuegbuetal(2011) studied polypropylene based composites with groundnut husk powder at varies particle size. The polypropylene composites were prepared in an extrusion moulding machine and the resulting composites were extruded as sheets. ThePresence of pulverized ground nut husk improved the tensile strength, modulus, flexural strength and impact strength of the composites and these properties are increased with increase in filler contents and decreases with filler particle size. Raju et al.(2012) investigated the groundnut shell powder reinforced polymer composite the composites were prepared with different weight% of particles in polymer matrix. From the experimental results. It has been observed that the sample with 20wt% of reinforcement, has maximum MOR (modulus of Rupture) of 40.57MPa and 60wt% reinforcement has maximum MOE(Modulus of Elasticity) of 8.204GPa. The impact test study showed increase in impact strength upto 50wt% of filler addition.

IOP Conf. Series: Materials Science and Engineering **574** (2019) 012023 doi:10.1088/1757-899X/574/1/012023

2. Materials and Method

In this current research work Epoxy resin(LY-556) was used as matrix material.HY951 was used as hardener for the prepration of composite. Chemically surface modified saw dust with jute fiber was used as reinforcement. Silicon spray was act as mould releasingagent.

Saw dust powder	Fabricated composite

Figure 1 Fabricated saw dust reinforced jute fabric composite

Hand layup process was used for fabricating the composite with the dimension of (300 × 300 × 3 mm) as per the Table 1.Figure 1 (a) shows the saw dust powder used for this investigation. Figure 1 (b) shows the fabricated saw dust reinforced jute fabric composite.

Table 1 Designation and composition of developed composites

Material code	**Filler (wt %)**	**Jute (wt %)**	**Epoxy (wt %)**
Saw dust 10 jute	10	40	50
Saw dust 20 jute	20	40	40
Saw dust 30 jute	30	40	30

3. Fracture Test (CT) Specimen Preparation

The fracture toughness was determined using Instron type universal testing machine as per ASTM D5045standard.The crack will begin on the point of the notch and extend through the sample. CT specimens are used extensively used in the area of testing, in order to establish fracture toughness values for a material. CT specimens are used for experiments where there is a shortage of material

IOP Conf. Series: Materials Science and Engineering **574** (2019) 012023 doi:10.1088/1757-899X/574/1/012023

available due to their compact design. Figure 1 shows test specimen used for this investigation. Stress intensity factor K can be considered as a estimate of fracture toughness. It depends on the applied load, flow depth and geometry of the specimen. Critical stress intensity factor for mode 1 can be determined from the following equation 1.

$K_{ic} = \frac{P}{B\sqrt{W}}\left\{F\left(\frac{A}{W}\right)\right\}$ ------------ Eq. 1

Where 'P' is the load at which crack propagate. 'B' denotes the thickness of the test specimen and 'w' is the length of the specimen. Crack length is 'a'.

$F\left(\frac{A}{W}\right)$- Geometry factor

$\left\{F\left(\frac{A}{W}\right)\right\} = \frac{2+\left(\frac{A}{W}\right)}{\left(1-\frac{A}{W}\right)^{1.5}}\left\{0.866 + 4.66\left(\frac{A}{W}\right) - 13.32\left(\frac{A}{W}\right)^2 + 14.72\left(\frac{A}{W}\right)^3 - 5.6\left(\frac{A}{W}\right)^4\right\}$------------ Eq. 2

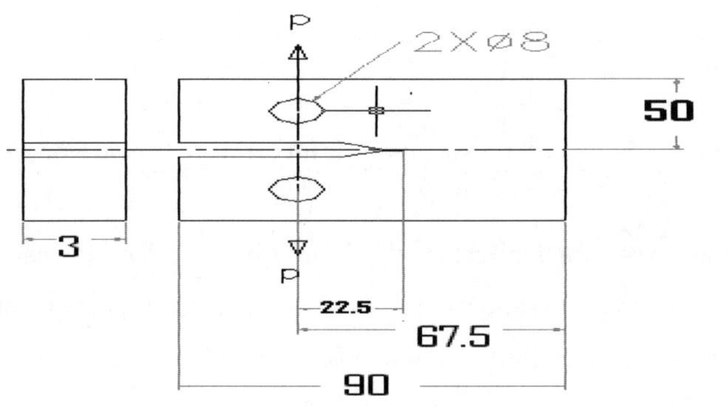

Figure: 2: Test specimen with specification

Universal testing machine is usedfor fracture test which is shown in Figure 2

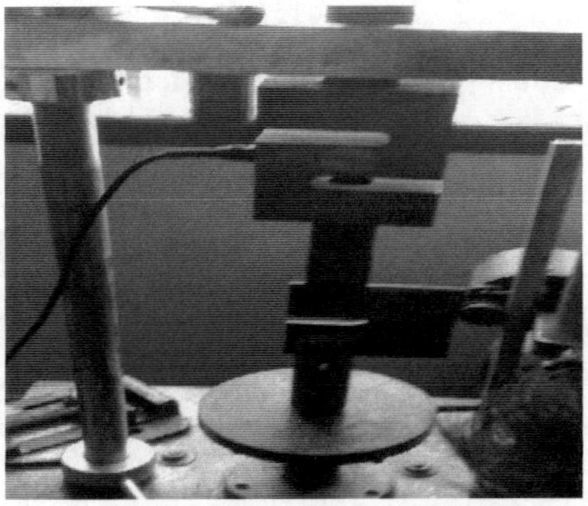

Fig.3 Testing of specimen

After finishing the fabrication of specimen, that specimen is fixed in the machine as shown in the Figure 3.The testing of the specimen is done with the help of universal testing machine. The 2 bolts are inserted in the 8mm diameter hole. The 2 bolts are fixed in the testing machine as shown in the Fig 3. Thus the specimen will be fixed in the testing machine for conducting CT test specimen. The testing machine will be connected with the computer to record the required results digitally. Fig 4. Shows the various specimens tested for this investigation.

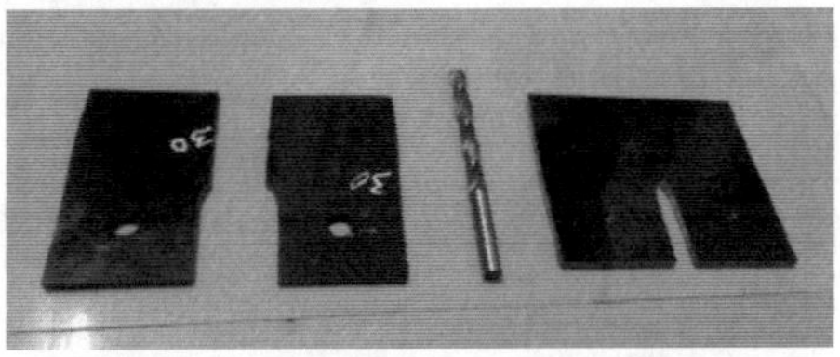

Fig.4 Specimens used for Mode I fracture

4. Result and discussion

Mode I fracture toughness was investigated using UTM machine. Table 2 shows the mode I fracture toughness value obtained from the experimental results.

Table 2 Mode I fracture toughness

Material code	Mode I fracture toughness	
	Untreated	**Treated**
Saw dust 10 jute	$1.2*10^{-3}$	$2.1*10^{-3}$
Saw dust 20 jute	$0.8*10^{-3}$	$1.9*10^{-3}$
Saw dust 30 jute	$1.8*10^{-3}$	$2.2*10^{-3}$

5. Mode I frature toughness

5.1 Effect of surface treated saw dust powder on mode I fracture toughness

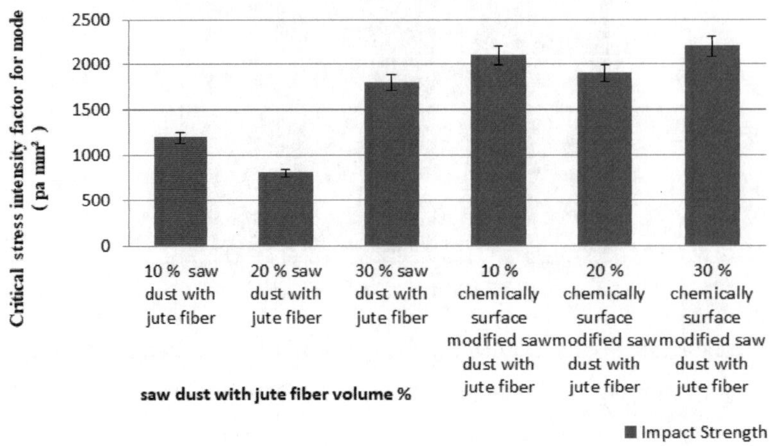

Figure 5 Effect of sawdust powder on mode I frature toughness

The effect of treated and untreated sawdust powder on fabricated composite mentioned in Figure 5. From the figure it was seen that increasing the w.t% of sawdust powder from 10% to 30% will increase the frature toughness. The uniformdistribution of sawdust particles in epoxy rein decrease the stress inrensity factor which in turn incresing the fracture toughness in the fabricated composite.

5.2. Microstructural morphology

The scanning electron microscope (SEM) wasused to identify the fracture morphology of the composite samples.

The surfaces of the composite specimens are examined directly by scanning electron microscope. The composite samples are washed, cleaned thoroughly and are coated with 120 Å thick platinum in sputter ion coater and observed SEM at 40 kV. Similarly the composite samples are mounted on stubs with silver paste. To enhance the conductivity of the samples, a thin film of platinum is vacuum-evaporated onto them before the photomicrographs are taken.

IOP Conf. Series: Materials Science and Engineering **574** (2019) 012023 doi:10.1088/1757-899X/574/1/012023

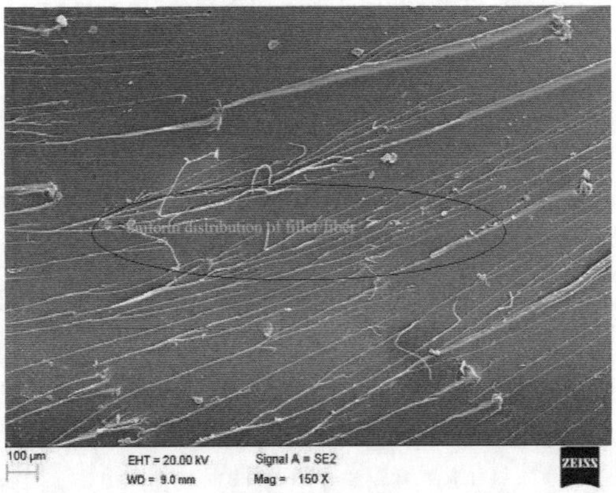

Fig 6 (a) Fractured specimen fiber with filler uniform distribution codition

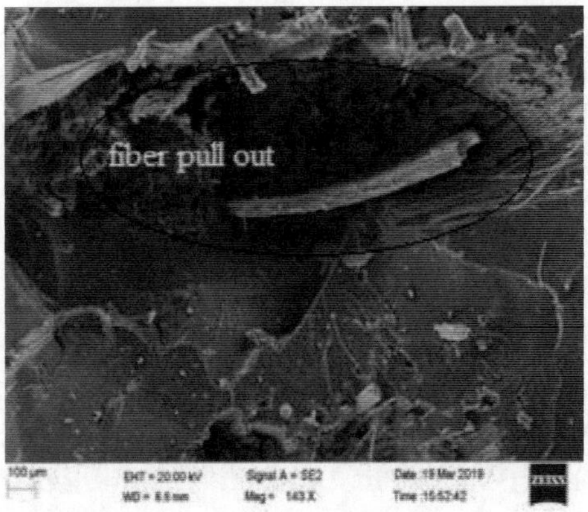

Fig 6 (b) Fractured specimen fiber pull out codition

Figure 6 SEM images of fractured specimen

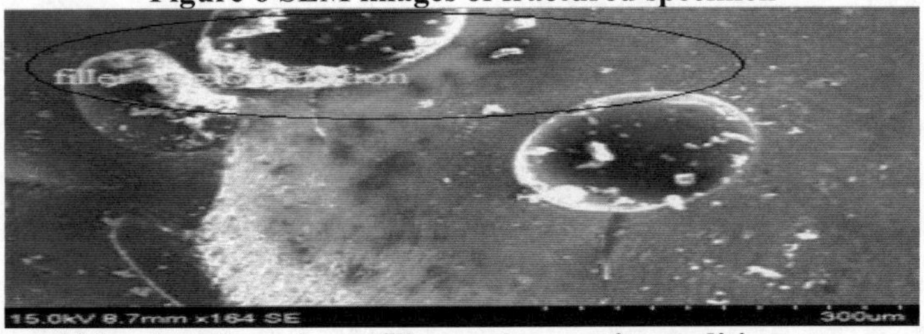

Fig 6 (c) SEM image of fractured specimen filler agglomeration codition

Figure 6 Scanning electron microscope image of fractured specimen

The test results were found out the value mentioned above Table 2. The maximum fracture properties were observed for the composite with 20 volume %chemically surface modified saw dust with jute fiber. The Increased with the incorporation of chemically surface modified saw dust with jute fiber this due to uniform distribution of saw dust with jute fiber in the epoxy matrix and Due to uniform dispersion, stress is created because of applied load is disseminated equally it is evident from SEM images Fig 6 (a) . It also indicates that bonding between treated filler/fiber and matrix is better when compared to the composite with another untreated filler/fiber material. 30volume % of filler that reveals clustering of saw dust in the epoxy composite it is evident from SEM image Fig 6 (b) and 6 (c).This agglomeration of filler material results in amorphous nature of the composite.

6. Conclusion

The influence of different volume fraction of malic acid treated saw dust with jute fiber reinforced filler/fiber addition in the epoxy resin on the mode-1 fracture where analyzed the fractured toughness results replaced that the treated 3-layer jute fiber with 20 % percentage filler in the epoxy matrix increase the mode-1 fractured toughness properties such as fracture of the composite material. Further additions of saw dust filler reduce the property due to the non-uniform dispersion in filler/fiber in the epoxy matrix. Comparison of the experiments results are filler/fiberadded un treated another composite. Hence saw dust/jute fiber is a one type of natural filler can also be used with epoxy resin.

REFERENCES

[1] Kokta, BV, Raj, RGDaneault, C 1989, 'Use of wood flour as filler in polypropylene: Studies on mechanical properties', Polym. Plast. Technol. Eng.,vol.28, no. 3, pp. 247–259.

[2] Sarki, J, Hassan, SB, Aigbodion, VS, Oghenevweta, JE 2011, 'Potential of using coconut shell particle fillers in eco-composite materials', Journal of Alloys and Compounds, vol.509,no.5, pp. 2381-2385.

[3] Chun, KS, Husseinsyah, S, Azizi, FN 2013, 'Characterization and properties of recycled polypropylene/coconut shell powder composites: Effect of sodium dodecyl sulfate modification', Polym. Plast. Technol. Eng., vol. 52, no. 3, pp. 287–294.

[4] Kuburi, LS, Dauda, M, Obada, DO, Umaru, S, Dodoo-Arhin, D, Iliyasu, I & Mustapha, S 2017, 'Effects of Coir Fiber Loading on the Physio-mechanical and Morphological Properties of Coconut Shell Powder Filled Low Density Polyethylene Composites', Procedia Manufacturing, vol. 7, pp. 138-144.

[5] Siró, I &Plackett, D 2010, 'Microfibrillated cellulose and new nanocomposite materials: a review', Cellulose, vol. 17, no. 3, pp. 459-494.

[6] Muthukumar, S &Lingadurai, K, 'Global Journal of Engineering Science and Researches'.

[7] Udhayasankar, R &Karthikeyan, B 2015, 'A Review on Coconut Shell Reinforced Composites', International Journal of Chem Tech Research., vol. 8, no. 11, pp. 624-637.

[8] Chandramohan, D & Kumar, AJP 2017, 'Experimental data on the properties of natural fiber particle reinforced polymer composite material', Data in brief, vol. 13, pp. 460-468.

[9] Onuegbu, GC &Igwe, IO 2011, 'The effects of filler contents and particle sizes on the mechanical and end-use properties of snail shell powder filled polypropylene', Materials Sciences and Applications, vol. 2, no. 07, p. 810.

[10] Raju, GU &Kumarappa, S 2011, 'Experimental study on mechanical properties of groundnut shell particle-reinforced epoxy composites', J. Reinf. Plast. Compos. vol. 30, no. 12, pp. 1029–1037.

Evaluation of Mechanical Characterization of CFRP Reinforced Stainless Steel (SS304) Wire Mesh Polymer Composite

S.D. Naga Kiran[1], V. Jayakumar[2], S. Madhu[3*]

[1]Department of Mechanical Engineering, Saveetha School of Engineering, Saveetha Institute of Medical and Technical Sciences, Chennai, 602105, Tamil Nadu, India.
[2]Department of Mechanical Engineering, Amrita School of Engineering, Amrita Vishwa Vidyapeetham, Chennai Campus, Chennai – 601 103, Tamil Nadu, India.
[3]Department of Automobile Engineering, Saveetha School of Engineering, Saveetha Institute of Medical and Technical Sciences, Chennai, 602105, Tamil Nadu, India
Corresponding Author: mathumarine@gmail.com

Abstract

In recent days, fibre metal hybrid composites have been utilized for improve the stiffness, excellent corrosion resistance, low density and less weight. CFRP reinforced stainless steel (SS) wire mesh hybrid fibre metal laminate has been fabricated using hand layup process. Viscoelastic properties of fabricated composite samples have been studied by using Dynamic Mechanical Analysis (DMA) test. Mechanical characterization such as Dynamic Mechanical Analysis (DMA), Flexural test and tensile properties of the newly fabricated fibre metal laminated also investigated. The DMA parameters such as storage modulus (E'), loss modulus (E'') and damping factor (tan δ) has been investigated. Flexural test has been performed for evaluating the bending stress.DMA results revealed that storage modulus of E' was maximum at 5 Hz.In this work tensile strength observed was 533.05 MPa at 16.19 KN. For Flexural strength obtained in the fabricated composite material was 1.03 KN.

Key words: Carbon fibre reinforced polymer, SS304 wire mesh, Epoxy Resin, DMA, Tensile, Flexural.

1. Introduction

Now a day's composite materials has been used in many areas such as manufacturing industries, automotive and aircraft industries. The primary reason of this composite material was chosen because of less weight, relative stiffness and high strength. Recent years have seen valuable research work on and widespread applications of fiber-reinforced polymer (FRP) bars as flexural reinforcement for concrete structures. CFRP reinforced stainless steel (SS) composite has excellent fatigue resistance and can increase product lifespan by several times in many applications such as bullet proof jacket and body parts of a car that can be used [1]. Recent advancements in metal fibres have introduced a promising new type of stainless steel fibre with high stiffness, high failure strain, and a thickness which can be utilized in a steel fiber reinforced polymer. However, stainless steel is known to be susceptible to pitting corrosion [2].Composites offer many advantages when compared to metallic alloys, especially where high strength and stiffness to weigh ratio is concerned. Additionally, they provide excellent fatigue properties and corrosion resistance in applications [3].Mechanical behaviours of pure titanium, new titanium alloy films and CFRP/Ti alloy laminated composites are investigated in order to research of development of fibre metal laminates tanks for cryogenic rocket propellants [4].

While conventional fiber-reinforced polymer composites offer high strength and stiffness, they lack ductility and the ability to absorb energy before failure. This work investigates hybrid fiber composites for structural applications comprised of polymer, steel fiber, and glass fibers to address this shortcoming. Varying volume fractions of thin, ductile steel fibers were introduced into glass fiber reinforced epoxy composites [5]. Impact of corrosion on mechanical properties of steel fiber reinforced composites was investigated. Corrosion of composite materials with plates of different types of stainless steel was compared. This work mentioned samples of steel reinforced polymer (SFRP) and four different types of stainless steel were subjected to 144 and 288 h of corrosion in ferric chloride solution [6].GFRP spirals are also efficient in providing shear strength, laterally supporting compression and tension bars, and confining the compressive concrete core. In the tested specimens, CFRP bars were used as longitudinal reinforcement, whereas GFRP spirals were used for transverse reinforcement [7].Tensile properties of a titanium-based fibre metal laminate have been investigated at quasi-static rates of strain. Initial attention focused on investigating the effect of varying the fibre orientation on the tensile modulus and strength of a range of carbon fibre/PEEK FMLs [8].

The influences of the processing parameters on the microstructure and mechanical properties of the stainless steel thin wall parts by laser direct deposition were investigated. The results show that the obtained microstructure of the stainless steel thin walls is all composed of directionally solidified dendrites under different processing parameters [9].This work reports on a study in which the flexural strength and deformability of circular concrete members with hybrid reinforcement CFRP bars and GFRP spirals. Test results indicate that the CFRP-RC specimens failed gradually at a high degree of deformability before concrete crushing [10]. Applicability of using CFRP bars as longitudinal reinforcement in circular concrete columns was investigated. Their test results indicated that the CFRP reinforced and steel reinforced columns tested behaved similarly up to their peak loads. The CFRP bars were effective in resisting compression until after concrete crushing and contributed to column capacity [11].This work revealed that, with the increasing of duration, corrosion potential of stainless steel after heat treatment shows fall followed by rise [12].Heat treatment is used to enhance the mechanical properties of the material. DMA are used to elucidate Tg and some mechanical properties like strength, stiffness and damping behaviour of the material [13].

In this work stainless steel (SS 304) reinforced carbon fibre reinforced fibre metal laminate has been fabricated using hand layup process. Mechanical catheterization such as DMA, flexural and tensile test was evaluated.

2. Materials and Method

Carbon fibre, Stainless Steel (SS304) wire mesh, epoxy resin and hardener were used for preparing the fibre metal composite. In this experimental work 5-10 μm in diameter carbon fibre has been used (Figure 1). SS304 stainless steel wire mesh with 50 microns is used for this investigation is shown in Figure 2. Mechanical properties of SS-304 are shown in Table 1.CFRP/ SS-304 composite is a fibre metal laminate is composed of thin stainless steel layer is inter pressed with four layers of carbon fibre is bonded with epoxy resin [14]. In this way four layers were arranged until the required thickness is achieved. The unidirectional layers of carbon fibre aligned in different directions to suit predicted conditions. 30N loads were applied on the laminates for 24hours to form a composite material is shown in Figure 3. The mould is allowed to cure for about 2-3 days for better results and later remove the composite from the mould and proceed for trimming or cutting as per required dimensions according to ASTM standards. The CFRP/ SS fibre metal laminate composite was fabricated as per the Table 1 and is shown in Figure 3.

Table 1 Mechanical Properties of SS 304 wire mesh

Properties	Value
Compressive Strength	310 MPa
Specific Heat	530 J/ Kg.K
Hardness	2100MPa
Tensile Strength	620 MPa
Young's Modulus	203 GPa
Melting Point	1723 K
Thermal Conductivity	17 W/ m.K

Figure 1 Carbon fibre used for this investigation

Figure 2 SS-304 wire mesh

Figure 3 Fabricated CFRP/SS-304

3. Mechanical Characterization

The mould is allowed to cure for about 2-3 days for better results and later remove the composite from the mould and proceed for trimming or cutting as per required dimensions according to ASTM standards

3.1 Dynamic Mechanical Analysis (DMA)

DMA is used to study the visco elastic behavior of CFRP-SS304 with temperature. The sample is subjected by a controlled stress. The test samples were cut into Length – 53 mm, Width- 10.2 mm and thickness- 2.3 mm as shown in Figure 4. The resultant sine wave response of the sample is measured using an LVDT sensor [15, 16]. The amplitude is used to calculate the storage modulus (E'). The phase lag is used to calculate the loss modulus (E"). The damping factor tan δ = E"/E'.In this investigation, fabricated composite is analysed under thermal and dynamic loading conditions using SEIKO DMAI-DMSC 6100. Dynamic Mechanical Analysis test was carried out at nitrogen environment with a temperature range of 25-200°C at 2°C/min increment with various frequencies of 0.2 Hz, 0.5 Hz, 1 Hz, 2 Hz and 5 Hz under the tensile mode.

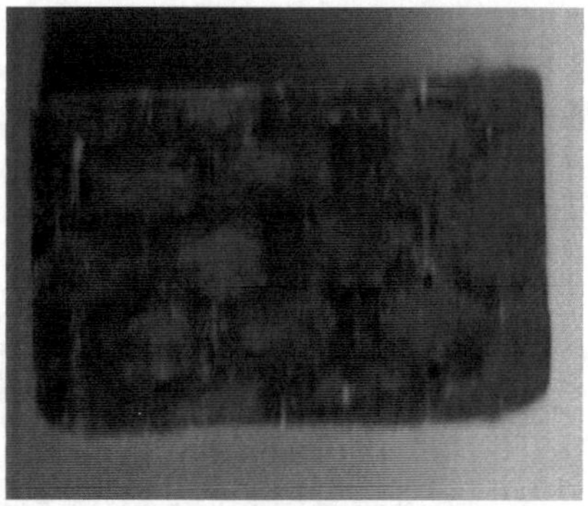

Figure 4 Test specimens for DMA

Figure 5 Tensile test specimen

Figure 6 Specimen for flexural test

3.2 Tensile Testing

The CFRP/SS-304 composite is performed in Universal Testing Machine (UTM) to investigate tensile load.The specimen of this composite is shown in Figure 5. The Testing material has been performed in ASTM D 638 [17]. The dimension of this specimen is 30.37mm ^{2} X 12.55 mm X 2.45 mm.

3.3 Flexural Test

This method used to measure the bending stress of a CFRP/SS-304 composite material under three point loading condition. The test specimen used for this investigation is shown in Figure 6. The sample is fixed under two point loading condition. The sample bended by the mandrel where force applied to the sample which is a third point loading in UTM machine.

4. Results and Discussion

4.1 Dynamic Mechanical Analysis (DMA)

4.1.1 Effect of temperature on Storage Modulus (E') for different frequencies

The storage modulus E' represents the stiffness of visco elastic behaviour of a sample material and it measure the stored energy during one loading cycle of the composite material CFRP/SS-304 of all frequencies. Figure7(a-e) explained the storage modulus of all frequencies 0.2 Hz, 0.5 Hz, 1 Hz, 2 Hz and 5 Hz.

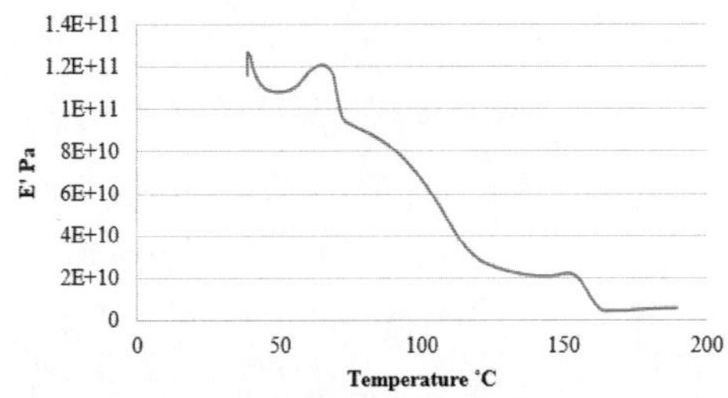

Figure 7(a) E' for 0.2 Hz

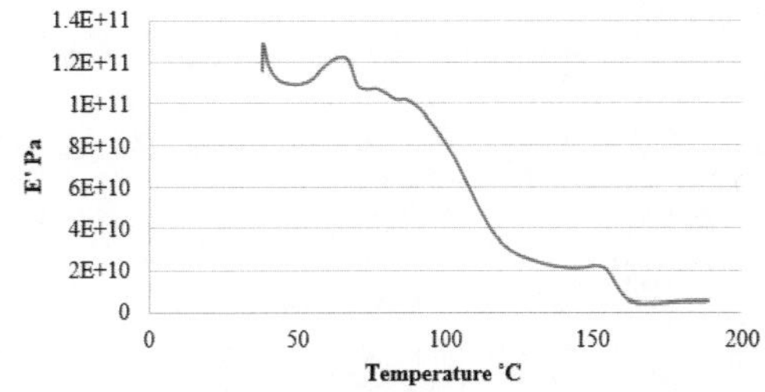

Figure 7(b) E' for 0.5 Hz

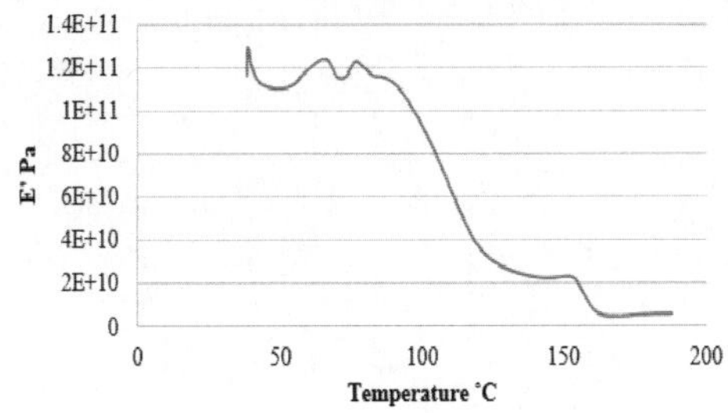

Figure 7(c) E' for 1 Hz

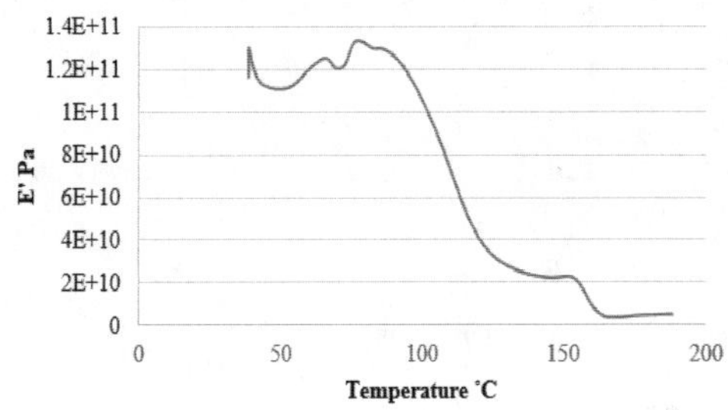

Figure 7(d) E' for 2 Hz

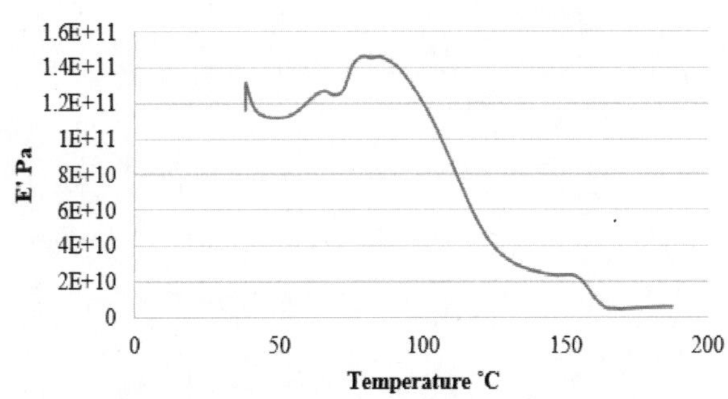

Figure 7(e) E' for 5 Hz

Figure 7 Storage Modulus (E')

Figure 7 shows the storage modulus (E') obtained from the DMA. From the above figure, it was observed that the frequency 5 Hz has high storage modulus at 85.17327 °C at 1.46E+11 Pa [18]. It was identified that the CFRP/SS-304 composite has high storage modulus at frequency 5Hz. After that the material loses its viscoelastic behaviour among the other frequencies.

4.1.2. Effect of temperature on Loss Modulus (E'') for different frequencies

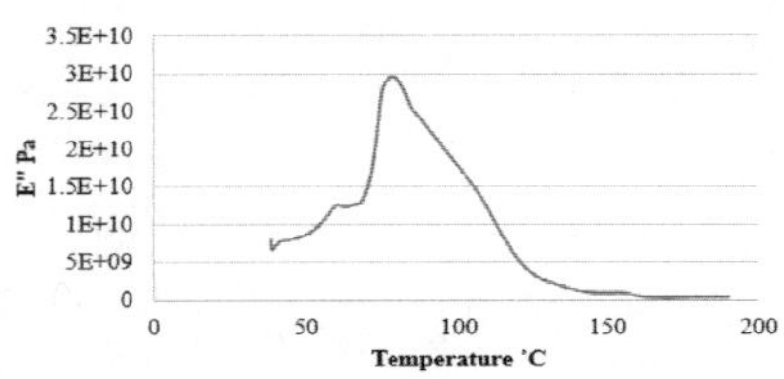

Figure 8(a) E'' for 0.2 Hz

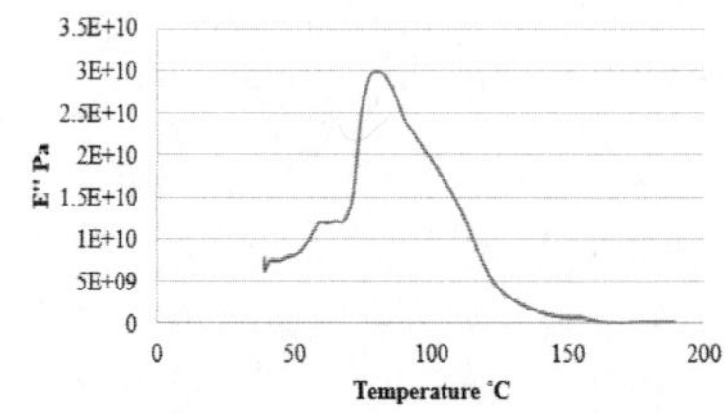

Figure 8(b) E'' for 0.5 Hz

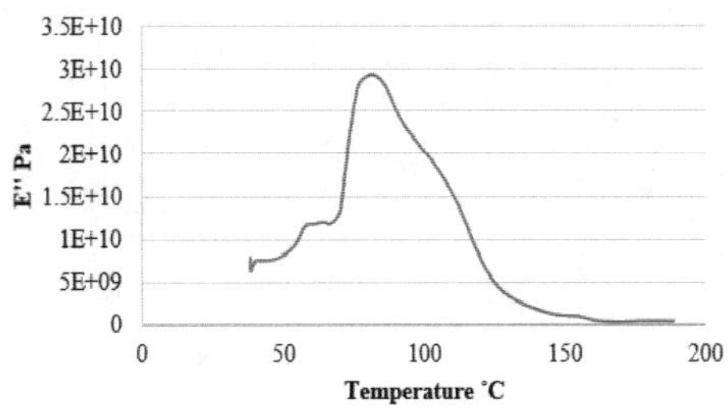

Figure 8(c) E'' for 1 Hz

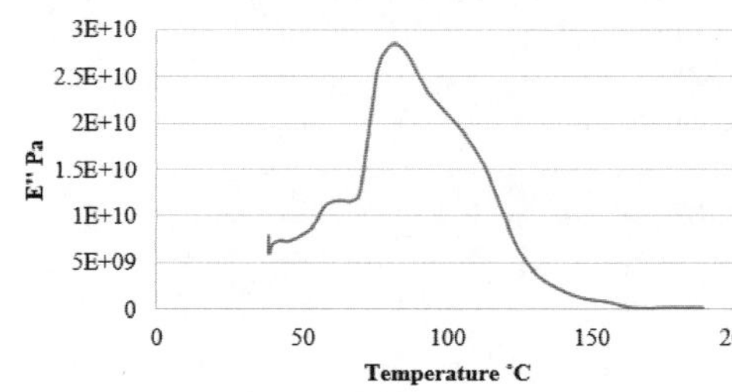

Figure 8(d) E'' for 2 Hz

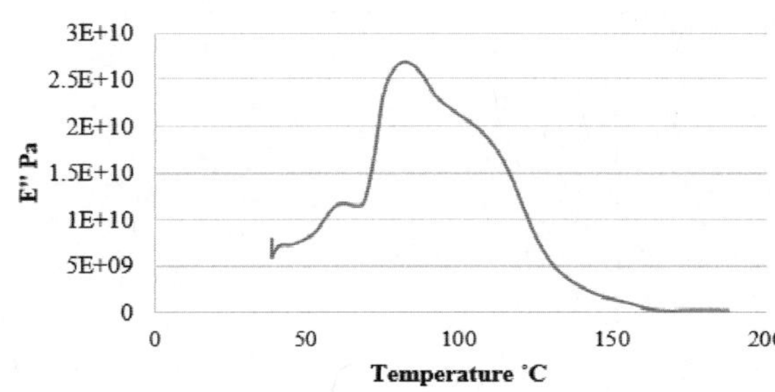

Figure 8(e) Temp v/s E'' for 5 Hz

Figure 8 Loss modulus (E")

The Loss Modulus E" represents the energy is dissipated during one loading cycle of the sample CFRP/SS-304 composite. Loss Modulus is decreases with increase in temperature by applying in the sample. Figure 8 (a-e) represents the results of loss modulus of all frequencies 0.2 Hz, 0.5 Hz, 1 Hz, 2 Hz and 5 Hz.

From the above figure 8, it was observed that frequency 0.5 Hz has high loss modulus (E") at 80.30649°C at 3E+10 Pa. It was identified that the CFRP/SS-304 composite has high loss modulus at 0.5 Hz. After that it loses the energy dissipated during one loading cycle of the sample CFRP/SS-304 [19].

4.1.3. Effect of temperature on Loss Factor (tan δ)for different frequencies

The loss factor tan δ is the ratio of loss modulus to storage modulus. It represents the mechanical damping and internal friction in a viscoelastic system of a given sample CFRP/SS-304

IOP Conf. Series: Materials Science and Engineering **574** (2019) 012024 doi:10.1088/1757-899X/574/1/012024

composite. Figure 9 (a-e) represents the graphs results of loss factor of all frequencies 0.2 Hz, 0.5 Hz, 1 Hz, 2 Hz and 5 Hz were drawn.

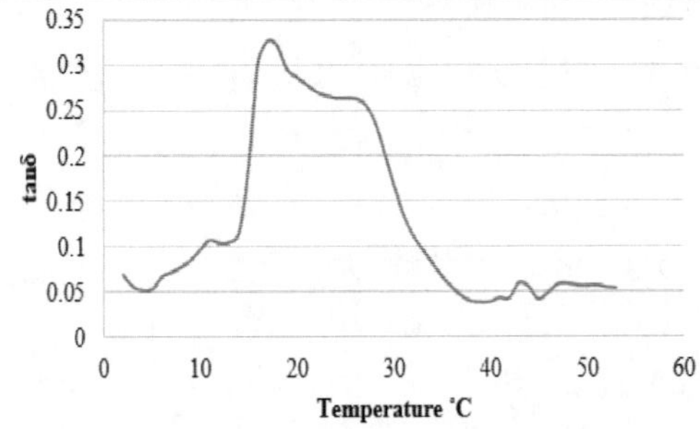

Figure 9(a) tan δ for 0.2 Hz

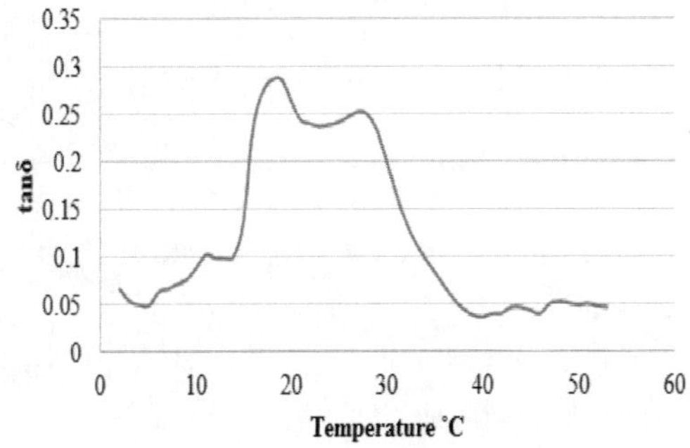

Figure 9(b) tan δ for 0.5 Hz

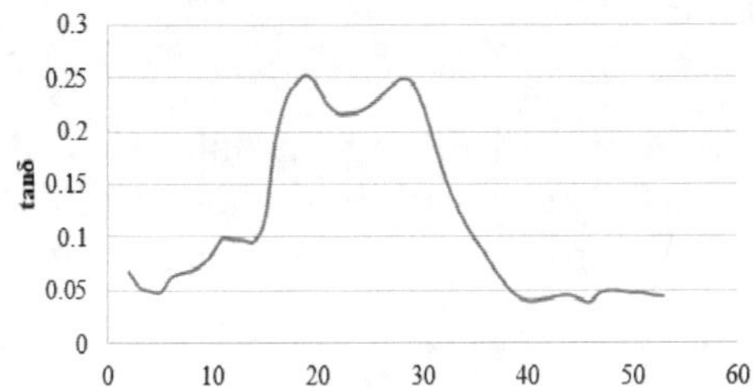

Figure 9(c) tan δ for 1 Hz

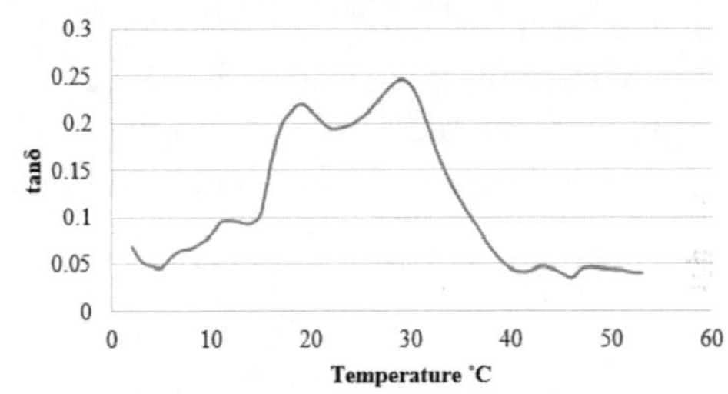

Figure 9(d) tan δ for 2 Hz

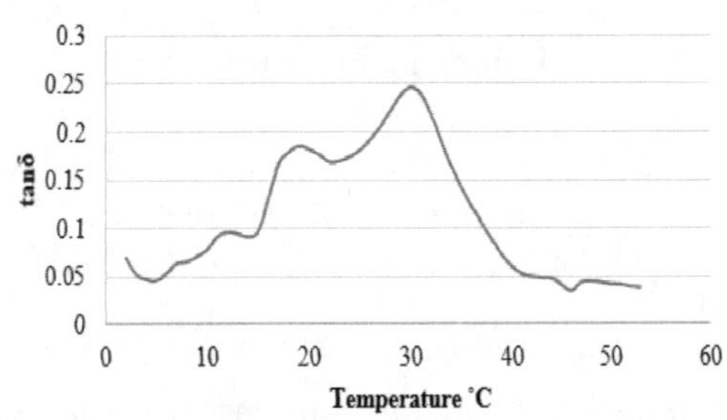

Figure 9(e) tan δ for 5 Hz

Figure 9 Loss factor (tan δ)

From the figure 9, it was observed that frequency 0.2 Hz has high loss factor (tanδ) at 78.07755 ˚C at 0.326386 loss factor [20]. It was identified that frequency 0.2 Hz has high loss factor compare to all other frequencies [22] of the given sample CFRP/SS-304 composite.

4.2 Tensile Test

The sample was test under Universal Testing Machine (UTM). While tensile test is applied to the sample, the grips are moves apart at a constant rate to pull and stretch the sample. The force of the sample and stroke is continuously monitored and plotted on stress-strain curve until sample failure or cracks are shown in Figure 10. The load displacement curve is plotted for determination of elastic modulus and ultimate tensile strength.

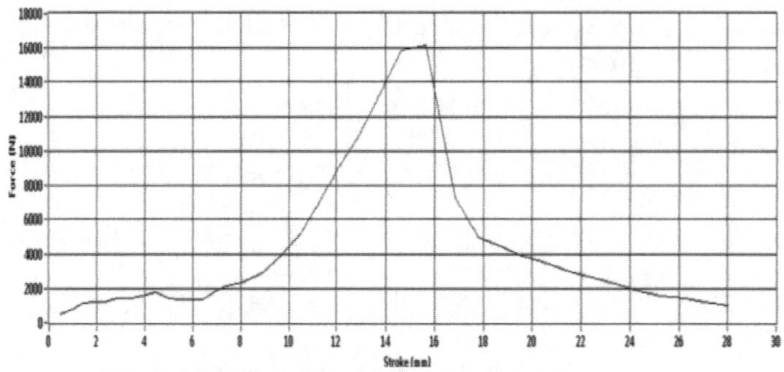

Figure 10 Ultimate Tensile strength

From figure 10, it was observed that the ultimate tensile strength of this sample CFRP/SS-304 composite is 533.05 MPa and maximum force where the sample breaks is 16.19 KN.

4.3 Flexural Test

The sample was tested for flexural strength under universal testing machine [21]. The results obtained in the form of graphical form from the UTM which is shown the Figure 11. During the bending stress, the change in the gauge section is recorded against the applied force.

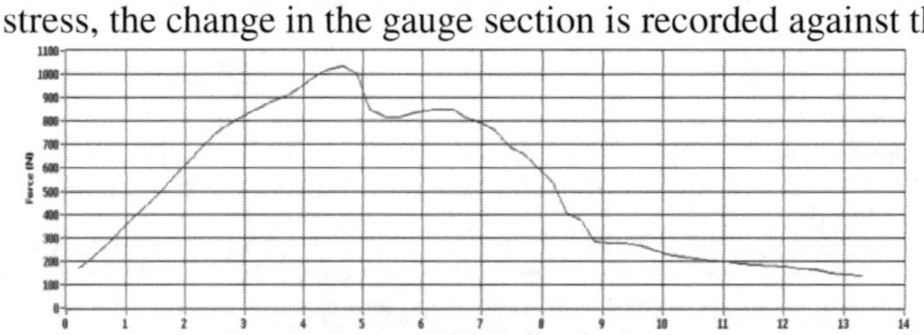

Figure 11 Flexural test

From the Figure, it was observed that the flexural load of this sample CFRP/SS-304 composite is N. The flexural strength is maximum compare to other natural fibres. For Flexural strength obtained in the fabricated composite material was 1.03 KN.

5. Conclusions

Carbon fibre reinforced SS fibre metal laminate was fabricated using handlyup process. Mechanical characterization such as DMA, tensile and flexural has been evaluated. This research work revealed that addition of stainless steel mesh in the carbon fibre composite increased the tensile strength. From the experimental results the following conclusions has been drawn

- The stainless steel wire mesh in the carbon fibre composite increase the tensile strength up to 533.05 MPa
- By substitute the stainless steel mesh with carbon fibers on the compressive surface yields the maximum flexural strength.
- DMA results revealed that storage modulus (E') obtained was maximum at 5 Hz

References

[1] Umer Mehmood, S.M.A.Saeed, A.Raza, M.Aziz and M.A.Naseer Hybridization Effect of Carbon Fibreon Mechanical Properties of Epoxy Based CompositesDepartment of Polymer & Process Engineering, University of Engineering and Technology, Lahore, 2007.

[2] M Devaiah, Comparison of mechanical properties of carbon, E-glass and hybrid composites,(IJMET) Volume 10, Issue 02, 2019, pp. 407-417.

[3] Jawed kadhim Uleiwi, Experimental study of Flexural strength of laminate composite material, Eng&Technology, Vol25, no.3, 2007, pp454-400.

[4] Amjad j. Aref and wael I, Alnahhal, Experimental Evolution of Hybrid FRP-Concrete Bridge SuperstructureSystem under negative Moment Flexural Loads, Jordon journal of Civil Engineering,Vol 1, No. 4, 2007, pp 336-343.

[5] Daniel I.M "New method for testing composites at very high strain rates",Experimental mechanics, Vol 21 No. 2 1981, pp 71 – 77.

[6] Hamilton W G, "Strain Rate characterization of Unidirectional Graphite/Epoxy composite, "Composite Materials Testing and Design

[7] Huiyong Ban and Mark A. Bradford "Flexural Behaviour of Composite Beams with High Strength Steel", Engineering Structures, Vol. 56, 1130-1141, 2013.

[8] Yaniv G, "Method for Hygromechanical Characterization of Graphite/Epoxy composite", Journal of Composites Technology & Research, vol 9, no.1 1987. pp 21-25.

[9] Murthy P L, Chamis C C, "Integrated Composite Analyzer for various tests", Journal of Composite Technology & Research, Vol 8 No.1,1986 pp 8 – 17.

[10] Dastin, S, Lubin G et.al, "Mechanical properties and test techniques for Reinforced plastic laminates, "Composite Materials Testing and Design, STP 460, ASTM 1969, pp 13 – 26.

Triantafillou, T.C. Gibson, L.J. (1987). Failure mode maps for foam core sandwich beams. Materials Sci. and Engineering, 95: 37–53.

[11] S.Shankar, experimental evaluation of flexural properties of polymer matrix composites.Volume 3, Issue 3, 2012, pp. 504-510.

[12] Najah R.Mohsin, "Static and Dynamic Analysis of Center Cracked Finite Plate Subjected To Uniform Tensile Stress Using Finite Element Method" International Journal of Mechanical Engineering &Technology (IJMET), Volume 6, Issue 1, 2015, pp. 56 – 70.

[13] S.Madhu, M. Balasubramanian (2018) "Effect of swirling abrasives induced by a novel threaded nozzle in machining of CFRP composites", The International Journal of Advanced Manufacturing Technology Volume 95, 9-12, pp 4175-4189.

[14] M. Sreenivasa Reddy, Soma. V. Chetty, Sudheer Premkumar and Reddappa H N, "Validation Of Hardness and Tensile Strength of Al 7075 Based Hybrid Composites Using Artificial Neural Networks" International Journal of Mechanical Engineering & Technology (IJMET), Volume 5, Issue 9, 2014, pp. 280 – 285.

[15] Baptista R., Mendao A., Guedes M. and Marat-Mendes R .An experimental study on mechanical properties of epoxy-matrix composites containing graphite filler. Procedia Structural Integrity. 1, 2016, pp. 074 – 081.

[16] Ali M. Synthesis and Characterization of Epoxy Matrix Composites Reinforced with Various Ratios of TiC. Jordan Journal of Mechanical and Industrial Engineering. 10, (4), 2016, pp. 231-237.

[17] Yuanxin Z., Farhana P., Lance L. Fracture and Fatigue behaviour of carbon nano fiber reinforced epoxy. Materials Science and Engineering: A, 452, 2007, pp. 657-664.

[18] Jenness J. R. Jr. and Kline D. E. Dynamic mechanical properties of some epoxy matrix composites. Journal of Applied Polymer Science. 17,(11), 1973, pp. 3391–3422.

[19] S.Y. Fua, B. Laukeb, E. Ma¨derb, C.Y. Yuea and X. Hua, Tensile properties of short-glassfiber- and short-carbon-fiber-reinforced polypropylene composites, Composites: Part A, Vol. 31, pp 1117–1125, 2000.

[20] Ching-Wen Lou, Kuo-Chen Chang, Chien-Teng Hsieh, Tung-Long Kuo, Jia-Horng Lin, Mechanical Analyses of Repeatedly Processed Polypropylene/Carbon Fiber Composite Plates, Advanced Materials Research Vols. 97-101, pp 1794-1796, 2010.

[21] Faria.R, Duncan. J.C, Brereton. R.G. Dynamic mechanical analysis and Chemo metrics for polymer identification. In: Polymer Testing, 26(3), 2007, pp: 403.